FORREST M. MIMMS III

ENGINEER'S MINI-NOTEBOOK SERIES

VOLUME II: SCIENCE AND COMMUNICATION CIRCUITS & PROJECTS

Published for Forrest M. Mims III by:
Master Publishing, Inc.
6019 W. Howard Street
Niles, IL 60714
847-763-0916 (voice)
847-763-0918 (fax)
masterpubl@aol.com (e-mail)

Visit Master Publishing on the Internet at:
www.masterpublishing.com

order on line at:
www.forrestmims.com

Printed in the United States of America

ABOUT THE ENGINEER'S MINI-NOTEBOOK COLLECTION

EACH BOOK IN THIS COLLECTION INCLUDES THREE OR FOUR ENGINEER'S MINI-NOTEBOOKS. EACH BOOK INCLUDES BOTH STANDARD CIRCUITS AND CIRCUITS DESIGNED BY FORREST M. MIMS III EACH CIRCUIT WAS BUILT AND TESTED AT LEAST TWICE. THE CIRCUITS WERE ALSO BUILT FROM THE FINAL BOOK TO FIND ERRORS.

VARIATIONS IN COMPONENTS AND CONSTRUCTION METHODS MAY CAUSE YOUR RESULTS TO DIFFER FROM THOSE DESCRIBED HERE. THEREFORE THE AUTHOR AND RADIOSHACK ARE NOT RESPONSIBLE FOR THE SUITABILITY OF THE CIRCUITS FOR ANY APPLICATION. FOR EXAMPLE, THE CIRCUITS IN THIS BOOK SHOULD NOT BE USED FOR MEDICAL APPLICATIONS, SAFETY DEVICES, TRAFFIC CONTROLLERS OR ANY OTHER USE THAT MIGHT SOMEHOW RESULT IN DAMAGE TO PROPERTY OR INJURY TO YOU OR OTHERS. IT IS YOUR RESPONSIBILITY TO DETERMINE IF COMMERCIAL USE, SALE OR MANUFACTURE OF ANY DEVICE BASED ON INFORMATION IN THIS BOOK INFRINGES ANY PATENT, COPYRIGHT OR OTHER RIGHT

FOR MORE INFORMATION

DUE TO THE MANY INQUIRIES RECEIVED BY THE AUTHOR AND RADIOSHACK, IT IS NOT POSSIBLE TO PROVIDE CUSTOM CIRCUIT DESIGNS AND TECHNICAL ADVICE YOU CAN LEARN MORE ABOUT ELECTRONICS FROM OTHER BOOKS AVAILABLE FROM RADIOSHACK AND FROM RADIOSHACK LAB KITS. ELECTRONICS MAGAZINES ARE ALSO A GOOD SOURCE OF INFORMATION. VARIOUS ELECTRONICS SITES ON THE INTERNET AND WORLD WIDE WEB ARE ALSO VERY HELPFUL.

II. ENVIRONMENTAL SCIENCE 53

UNITS OF MEASUREMENT

THE METRIC SYSTEM IS USED ALMOST EXCLUSIVELY IN SCIENCE. PRINCIPLE UNITS IN <u>THIS</u> BOOK:

INCHES TO MILLIMETERS = INCHES × 25.4
MILLIMETERS TO INCHES = MILLIMETERS × 0.03937
INCHES TO CENTIMETERS = INCHES × 2.54
CENTIMETERS TO INCHES = CENTIMETERS × 0.3937
FEET TO METERS = FEET × 0.3048
METERS TO FEET = METERS × 3.281
YARDS TO METERS = YARDS × 0.9144
METERS TO YARDS = METERS × 1.094
MILES TO KILOMETERS = MILES × 1.609
KILOMETERS TO MILES = KILOMETERS × 0.6214

TEMPERATURE — THE CELSIUS SCALE IS USUALLY USED IN SCIENCE. WATER FREEZES AT 0°C AND BOILS AT 100°C (SEA LEVEL). ROOM TEMPERATURE IS AROUND 23°C.
FAHRENHEIT TO CELSIUS = (°F − 32) × 5/9
CELSIUS TO FAHRENHEIT = (°C × 9/5) + 32

RADIO COMMUNICATIONS 124

HISTORICAL MILESTONES

1836 – SAMUEL F. B. MORSE INVENTS TELEGRAPH.
1876 – ALEXANDER GRAHAM BELL INVENTS TELEPHONE.
1880 – ALEXANDER GRAHAM BELL INVENTS PHOTOPHONE.
1880 – PHOTOPHONE SENDS VOICE 213 METERS.
1886 – HEINRICH HERTZ INVENTS SPARK TRANSMITTER.
1895 – GUGLIELMO MARCONI INVENTS WIRELESS TELEGRAPH.
1897 – NIKOLA TESLA SENDS RADIO SIGNAL 20 MILES.
1899 – MARCONI SENDS "..." ACROSS ATLANTIC OCEAN.
1899 – A. FREDERICK COLLINS SENDS VOICE OVER RADIO.
1907 – LEE DE FOREST INVENTS TRIODE VACUUM TUBE.
1907 – H. J. ROUND DISCOVERS LIGHT EMITTING DIODE.
1923 – O. V. LOSSEV INVENTS CRYSTAL AMPLIFIERS.
1925 – J. E. LILIENFELD INVENTS FIELD-EFFECT AMPLIFIER.
1947 – BELL LABS INVENTS TRANSISTOR
1960 – T. H. MAIMAN BUILDS FIRST RUBY LASER.
1962 – G.E., MIT AND IBM INVENT SEMICONDUCTOR LASER.
1966 – K.C. KAO PROPOSES OPTICAL FIBERS FOR LONG DISTANCE LIGHTWAVE LINKS.

I. SCIENCE PROJECTS

OVERVIEW

SCIENCE IS KNOWLEDGE GAINED BY ORGANIZED OBSERVATION, EXPERIMENTATION AND STUDY. THE PROJECTS THAT FOLLOW DEMONSTRATE BASIC SCIENTIFIC PRINCIPLES AND TECHNIQUES. SOME WILL LET YOU MEASURE TEMPERATURE, WIND SPEED, LIGHT AND POSITION. OTHERS WILL LET YOU DETECT RAIN, MOTION AND EARTH MOVEMENTS. YOU CAN LEARN MUCH BY TRYING ANY OF THE PROJECTS. YOU CAN LEARN EVEN MORE BY COMBINING PROJECTS AND MODIFYING THEM FOR OTHER PURPOSES. HERE ARE SOME SUGGESTIONS.

1. PLAN YOUR PROJECTS. DECIDE WHAT YOU WANT TO BUILD, MEASURE OR DETECT. SET GOALS AND ACCOMPLISH THEM.

2. KEEP A NOTEBOOK RECORD YOUR CIRCUITS, MEASUREMENTS AND OBSERVATIONS. BE ACCURATE. SIGN AND DATE EVERY PAGE. (THIS SERIES OF BOOKS EVOLVED FROM THE AUTHOR'S LAB NOTEBOOKS)

3 EXPERIMENT. FOR INSTANCE, SUBSTITUTE A LIGHT SENSOR FOR A THERMISTOR TO MEASURE LIGHT INSTEAD OF TEMPERATURE.

4. WANT TO KNOW MORE ABOUT A TOPIC? READ OTHER BOOKS IN THIS SERIES. VISIT A LIBRARY. READ ELECTRONICS MAGAZINES.

<u>SPECIAL NOTE TO STUDENTS, PARENTS AND TEACHERS</u>: MANY OF THE PROJECTS THAT FOLLOW CAN BE USED IN SCIENCE FAIR PROJECTS. FOR EXAMPLE, MEASURE BOTH TEMPERATURE AND SUNLIGHT ON CLEAR, PARTLY CLOUDY AND OVERCAST DAYS. GRAPH THE RESULTS. TEST THE INVERSE SQUARE LAW (PP.22-23) WITH VARIOUS LIGHT SOURCES. GRAPH THE RESULTS

ELECTROSCOPE

THE ELECTROSCOPE IS A SIMPLE DEVICE THAT WILL DETECT AN ELECTROSTATIC CHARGE AND THE PRESENCE OF NUCLEAR RADIATION. YOU CAN ASSEMBLE AN ELECTROSCOPE FROM COMMON HOUSEHOLD MATERIALS. FOR EXAMPLE:

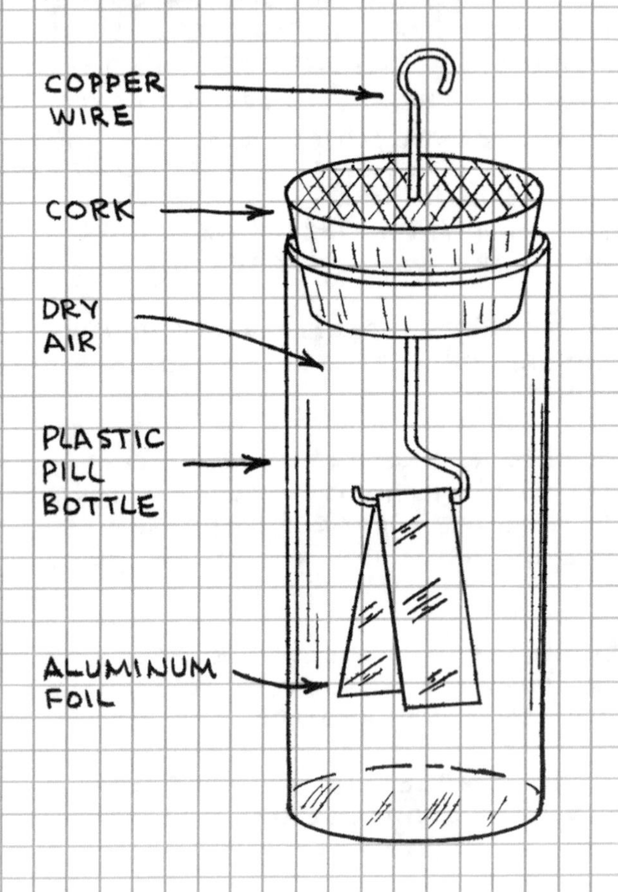

YOU CAN USE MANY DIFFERENT BOTTLES. THE BOTTLE MUST BE GLASS OR PLASTIC. THE STOPPER MUST BE CORK OR PLASTIC BUT NOT METAL. THE FOIL SHOULD BE THIN GAUGE ALUMINUM FOIL. THE AIR IN THE BOTTLE SHOULD BE AS DRY AS POSSIBLE.

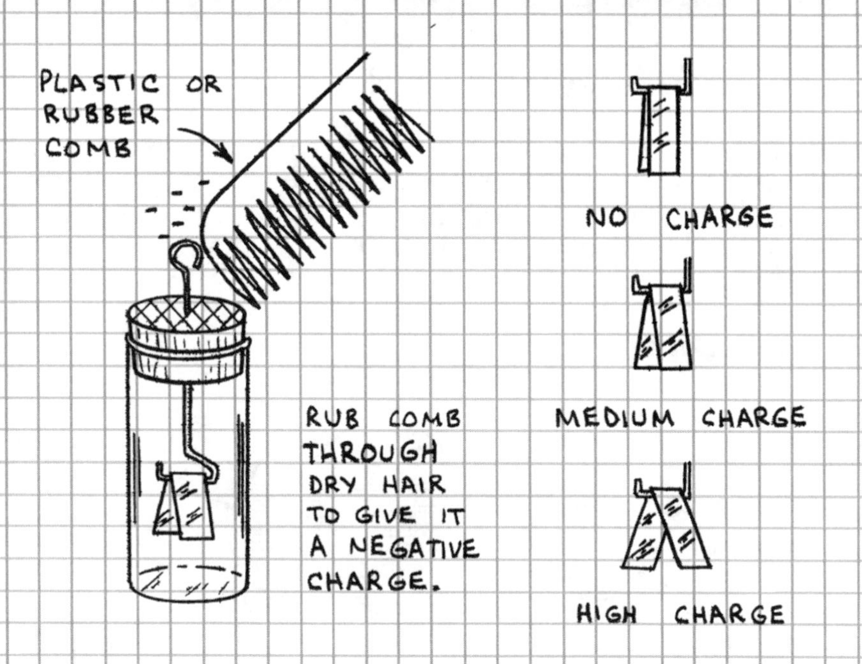

FOR BEST RESULTS THE FOIL LEAVES OF THE ELECTROSCOPE SHOULD BE FLAT. CUT THE FOIL WITH SHARP SCISSORS TO AVOID FRAYED EDGES. IF THE LEAVES DO NOT FLY APART WHEN A CHARGED OBJECT IS TOUCHED TO THE ELECTRODE, CHECK TO SEE IF THE LEAVES ARE STUCK TOGETHER. WORKS BEST WHEN AIR IS DRY. RADIATION WILL IONIZE THE AIR AND CAUSE LEAVES TO COLLAPSE.

ELECTRONIC ELECTROSCOPE

NORMALLY LED GLOWS. RUB PLASTIC COMB OR PEN THROUGH DRY HAIR AND PLACE CHARGED COMB OR PEN NEAR ELECTRODE. LED WILL BE EXTINGUISHED.

Q1 - USE 2N3819 OR SIMILAR N-FET.

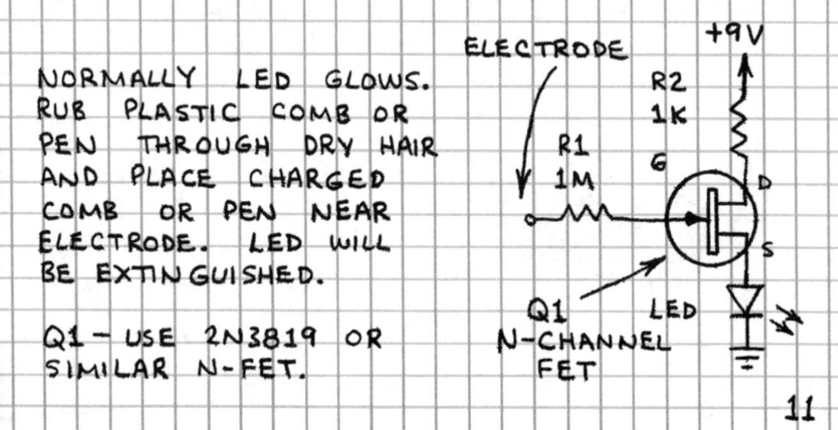

GALVANOMETER

THE GALVANOMETER MEASURES THE FLOW OF AN ELECTRICAL CURRENT. THE SIMPLEST GALVANOMETER IS MADE BY WRAPPING A WIRE COIL AROUND A COMPASS:

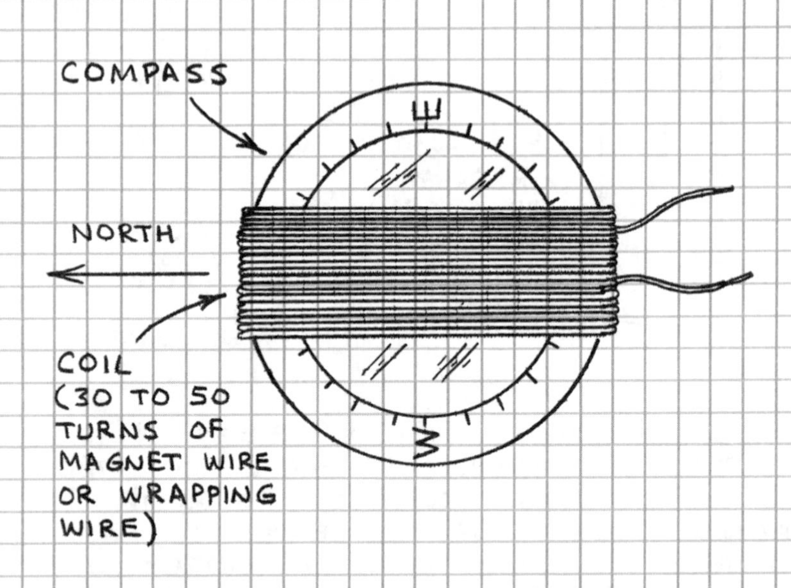

USE TAPE OR HOT MELT GLUE TO HOLD COIL IN PLACE. PLACE GALVANOMETER ON FLAT SURFACE. ALIGN SO THAT COIL AND COMPASS NEEDLE BOTH POINT NORTH. THEN TOUCH THE LEADS FROM THE COIL TO THE ENDS OF A 1.5 VOLT CELL. THE COMPASS NEEDLE WILL IMMEDIATELY SWING TO AN EAST-WEST ORIENTATION.

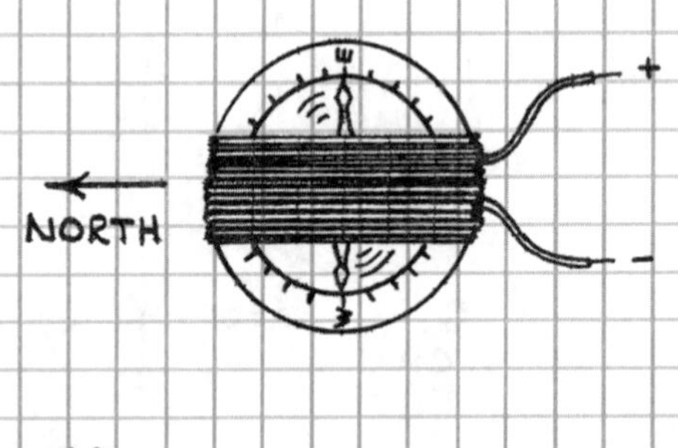

REVERSE POLARITY OF BATTERY TO REVERSE DIRECTION THE NEEDLE SWINGS. MOMENTARY USE WILL PREVENT EXCESS CURRENT DRAIN.

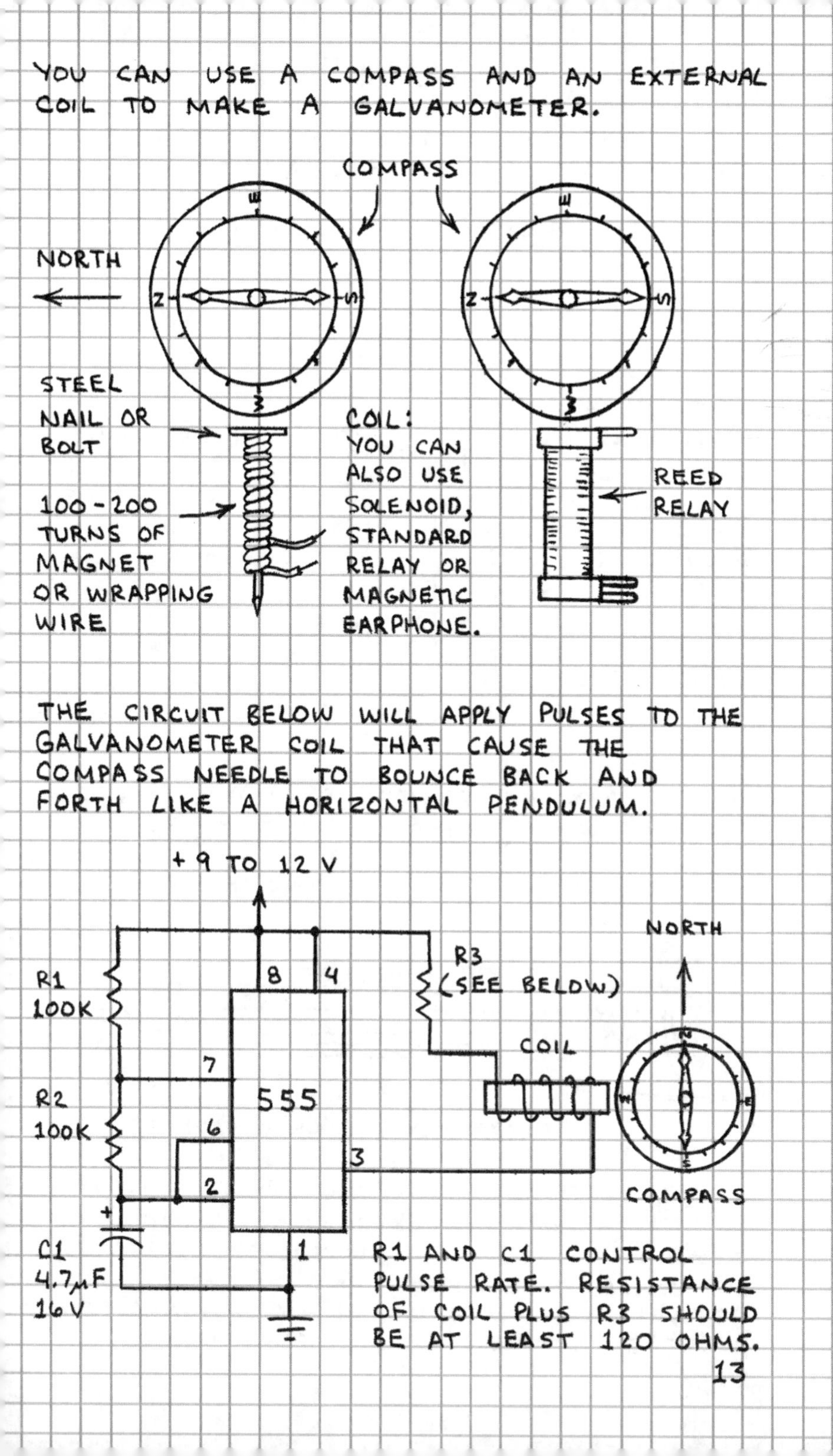
YOU CAN USE A COMPASS AND AN EXTERNAL COIL TO MAKE A GALVANOMETER.
COMPASS
NORTH
N
S
E
STEEL NAIL OR BOLT
100-200 TURNS OF MAGNET OR WRAPPING WIRE
COIL: YOU CAN ALSO USE SOLENOID, STANDARD RELAY OR MAGNETIC EARPHONE.
REED RELAY
THE CIRCUIT BELOW WILL APPLY PULSES TO THE GALVANOMETER COIL THAT CAUSE THE COMPASS NEEDLE TO BOUNCE BACK AND FORTH LIKE A HORIZONTAL PENDULUM.
+9 TO 12 V
R1 100K
R2 100K
C1 4.7µF 16V
555
8
4
7
6
2
3
1
R3 (SEE BELOW)
COIL
NORTH
COMPASS
R1 AND C1 CONTROL PULSE RATE. RESISTANCE OF COIL PLUS R3 SHOULD BE AT LEAST 120 OHMS.
13

HOMEMADE BATTERIES

HOMEMADE POWER CELLS AND BATTERIES CAN BE USED TO OPERATE MANY KINDS OF LOW POWER CIRCUITS. A BASIC CELL INCLUDES THESE COMPONENTS:

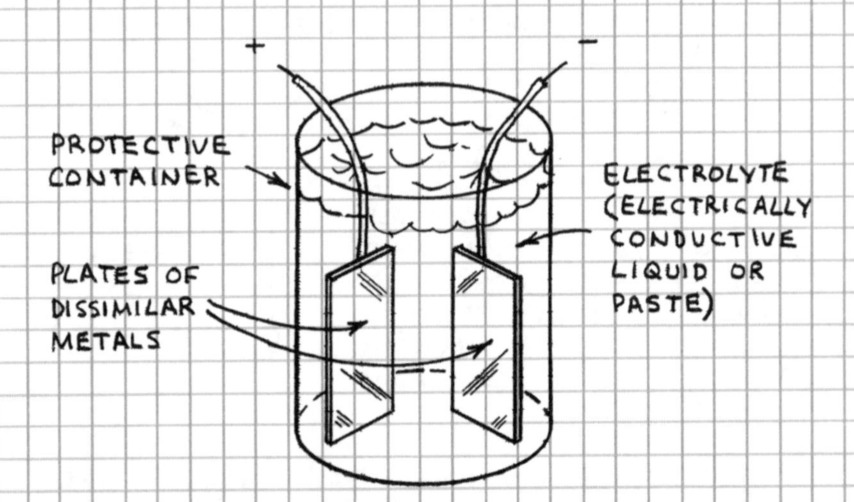

THERE ARE MANY WAYS TO MAKE PRACTICAL POWER CELLS. HERE IS AN EXAMPLE:

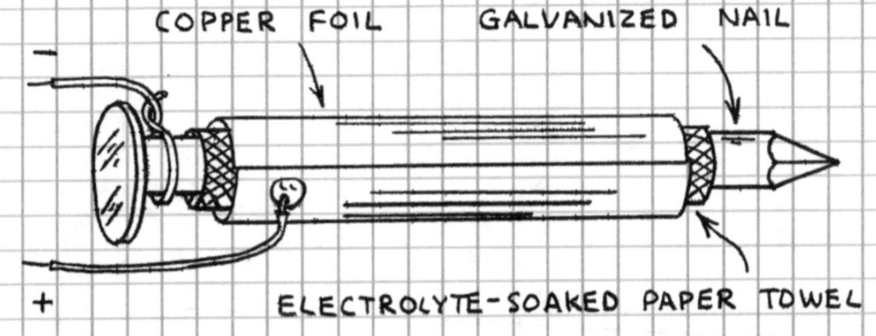

ELECTROLYTE CAN BE TABLE SALT DISSOLVED IN WATER OR POWDERED CITRIC DRINK (MUST CONTAIN CITRIC ACID) DISSOLVED IN WATER. DIP TOWEL IN SOLUTION AND ALLOW TO DRY. ACTIVATE CELL WITH WATER. CLEAN AND REUSE ELECTRODES WHEN CELL STOPS WORKING.

VOLTAGES MEASURED WITH VARIOUS ELECTRODE METALS AND ELECTROLYTES:

	ELECTRODES		ELECTROLYTE SALT	ACID*
1.	COPPER (+)	ZINC (-)	0.759	1.000
2.	COPPER (-)	SILVER (+)	0.200	0.131
3	COPPER (+)	MAGNESIUM (-)	1.400	1.484
4	COPPER (+)	ALUMINUM (-)	0.570	0.720
5.	ZINC (-)	SILVER (+)	0.720	0.820
6.	ZINC (+)	MAGNESIUM (-)	0.622	0.546
7.	ZINC (-)	ALUMINUM (+)	0.248	0.350
8.	ALUMINUM (+)	MAGNESIUM (-)	0.778	0.820
9.	ALUMINUM (-)	SILVER (+)	0.395	0.450
10.	SILVER (+)	MAGNESIUM (-)	1.242	1.231

* POWDERED CITRIC DRINK IN WATER.

WHERE TO FIND ELECTRODE MATERIALS:

COPPER — COPPER FOIL FROM A HOBBY SHOP OR COPPER LAMINATED CIRCUIT BOARD.

ZINC — GALVANIZED METAL AND NAILS FROM A HARDWARE STORE.

ALUMINUM — HOUSEHOLD ALUMINUM FOIL OR THIN SHEET ALUMINUM FROM A HOBBY SHOP.

SILVER — SILVER COIN OR THIN SILVER SHEET FROM JEWELRY SUPPLY STORE.

MAGNESIUM — THIN MAGNESIUM RIBBON FROM CHEMICAL SUPPLY COMPANY OR HOBBY SHOP.

THE VOLTAGES GIVEN IN THE TABLE ABOVE WERE MEASURED WITH A DIGITAL VOLTMETER. IN MOST CASES THE VOLTAGE BEGAN TO DECLINE ALMOST IMMEDIATELY. IN SOME CASES THE VOLTAGE INCREASED TO TWICE ITS INITIAL VALUE AFTER 20 SECONDS OR SO. PEAK VALUES ARE GIVEN IN EACH CASE

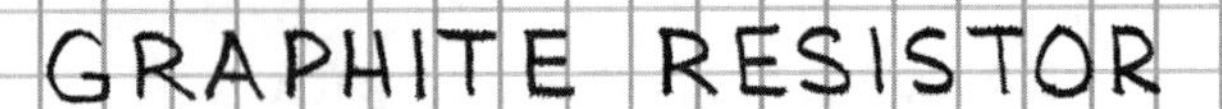

GRAPHITE RESISTOR

RESISTORS RESIST THE FLOW OF ELECTRICAL CURRENT. YOU CAN MAKE A RESISTOR BY STROKING A GRAPHITE PENCIL ON PAPER.

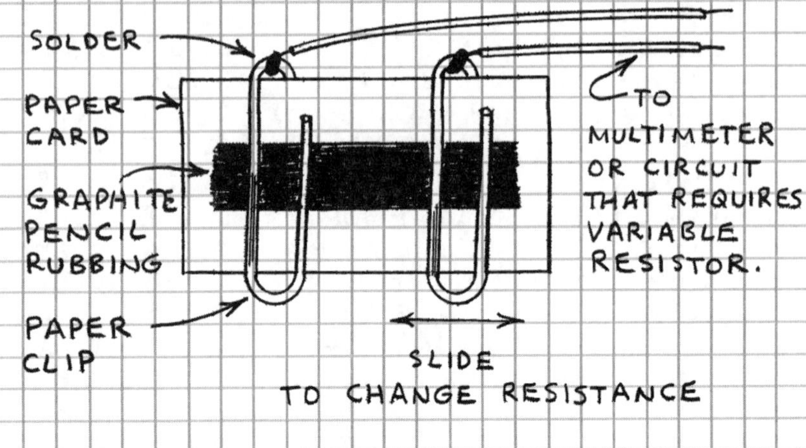

LIQUID RESISTOR

HERE'S HOW TO MAKE A RESISTOR FROM A CONDUCTIVE LIQUID (ELECTROLYTE).

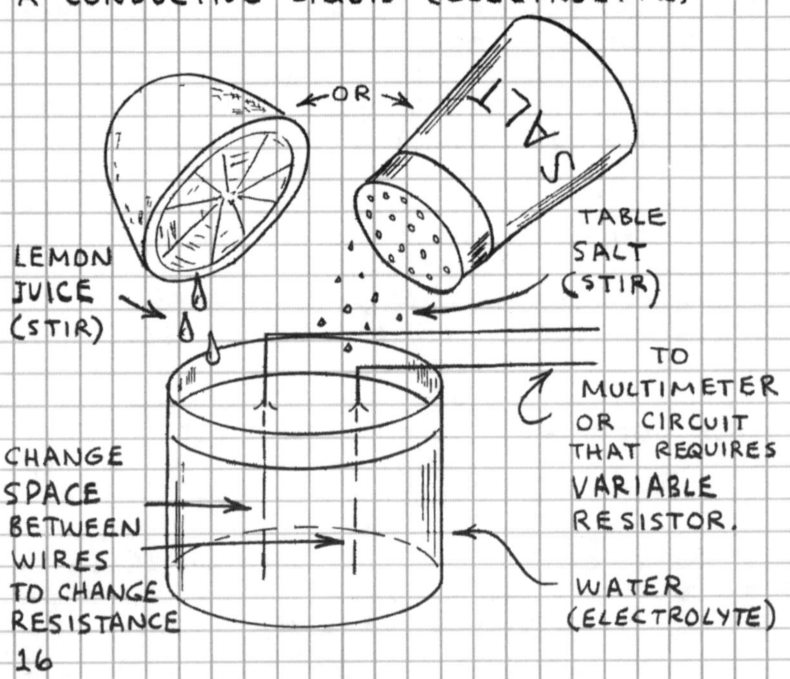

SUPER CAPACITOR

SUPER CAPACITORS STORE CONSIDERABLY MORE ENERGY THAN ORDINARY CAPACITORS. HERE'S HOW TO MAKE ONE:

COPPER-CLAD PC BOARD (FOIL SIDE DOWN)

1"

💧 = LEMON JUICE

+ OR −

💧💧💧 ACTIVATED CARBON FILTER *

💧💧💧 PAPER TOWEL

− OR +

COPPER-CLAD PC BOARD (FOIL SIDE UP)

* SOLD IN SHEETS AT PET AND AQUARIUM STORES.

USE RUBBER BAND TO HOLD CAPACITOR TOGETHER. THEN SOAK CARBON FILTER AND PAPER TOWEL LAYERS IN LEMON JUICE (ELECTROLYTE). INCREASE AREA FOR MORE CAPACITY. ADD LAYERS TO INCREASE VOLTAGE (1.2 VOLTS PER LAYER). DO NOT APPLY MORE THAN 1.2 VOLTS PER LAYER OR THE ELECTROLYTE WILL DECOMPOSE.

CHARGE (C) THROUGH 1K

SELF DISCHARGE

DISCHARGE (D) THROUGH LED AND 680 Ω

VOLTS: 1, 2, 3, 4

TIME (MINUTES): 0, .5, 1.0, 1.5, 2.0, 2.5

C, D, 1K, 680, 6V, C1, LED, +, −

PLACE C1 IN DISH.

THERMOCOUPLE

A THERMOCOUPLE IS MADE BY CONNECTING A WIRE OF ONE METAL BETWEEN TWO WIRES OF A SECOND METAL. IF ONE OF THE TWO CONNECTIONS OR JUNCTIONS IS MADE WARMER THAN THE OTHER, THEN THE THERMOCOUPLE WILL GENERATE A SMALL VOLTAGE. SOME METALS AND ALLOYS WORK MUCH BETTER THAN OTHERS IN THERMOCOUPLES. YOU CAN MAKE A SIMPLE THERMOCOUPLE FROM A PAPER CLIP AND SOME COPPER WIRE:

COLD (REFERENCE) JUNCTION

HOT JUNCTION

SOLDER HERE

COPPER

SOLDER HERE

PAPER CLIP WIRE

CONNECT THIS SIMPLE THERMOCOUPLE TO A DIGITAL MULTIMETER. IT WILL GENERATE UP TO A MILLIVOLT OR SO (0.001 VOLT) WHEN THE HOT JUNCTION IS HEATED BY A MATCH.

A <u>THERMOPILE</u> IS A SERIES OF MANY THERMOCOUPLES THAT GENERATES MORE VOLTAGE THAN A SINGLE THERMOCOUPLE:

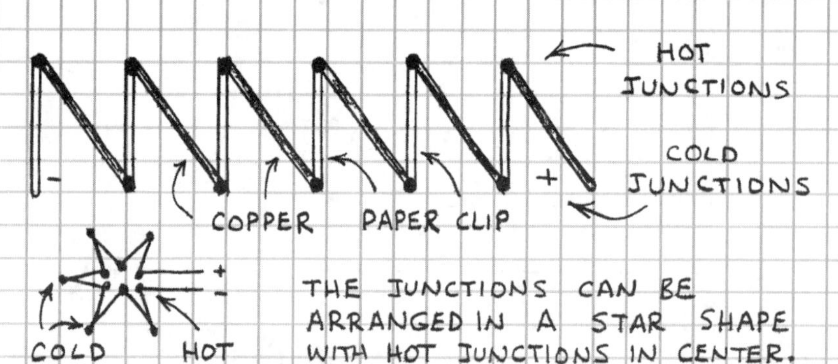

THE JUNCTIONS CAN BE ARRANGED IN A STAR SHAPE WITH HOT JUNCTIONS IN CENTER.

THERMOCOUPLE AMPLIFIER

AN OPERATIONAL AMPLIFIER WILL AMPLIFY THE TINY VOLTAGE GENERATED BY A THERMOCOUPLE.

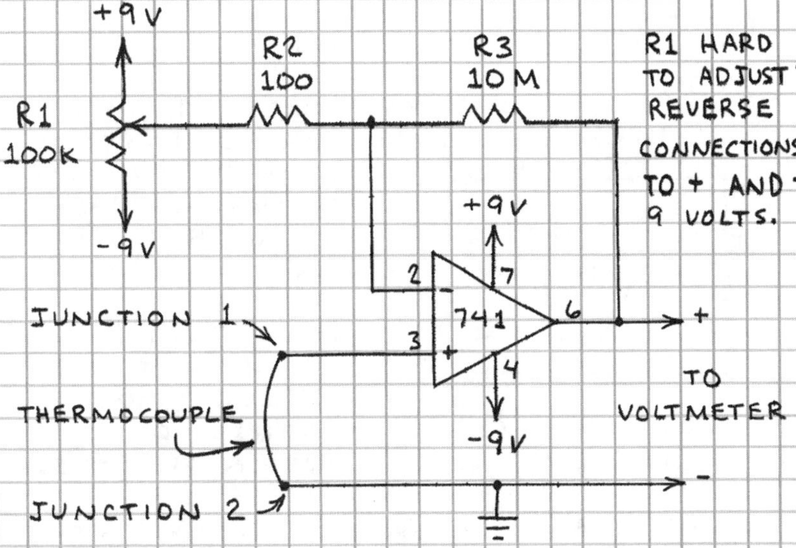

+9V
R1
100K
-9V
R2
100
R3
10 M
+9V
7
2
741
6
3
4
-9V
JUNCTION 1
THERMOCOUPLE
JUNCTION 2
+
TO
VOLTMETER
-

R1 HARD TO ADJUST? REVERSE CONNECTIONS TO + AND - 9 VOLTS.

9V
9V
+9V
-9V

R1 CONTROLS OFFSET VOLTAGE.

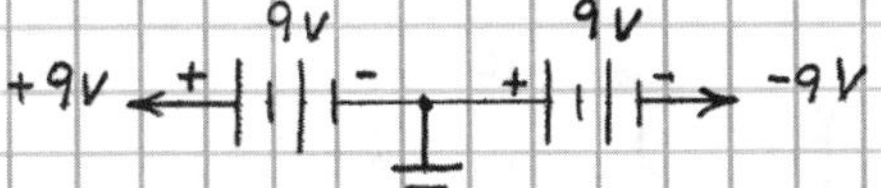

THE OUTPUT VOLTAGE FALLS WHEN JUNCTION 1 IS WARMER THAN JUNCTION 2. THE OUTPUT VOLTAGE RISES WHEN JUNCTION 2 IS WARMER THAN JUNCTION 1. FOR BEST RESULTS, USE ANALOG VOLTMETER FIRST. SET R1 FOR OUTPUT OF A FEW TENTHS OF A VOLT. THIS WILL LET YOU WATCH OUTPUT VOLTAGE SWING BACK AND FORTH, DEPENDING ON WHICH JUNCTION IS WARM. AFTER YOU LEARN TO ADJUST R1 (BE PATIENT) YOU CAN USE A DIGITAL VOLTMETER. NOTE THAT THE RISING OR FALLING VOLTAGE CAUSED BY HEATING ONE OF THE JUNCTIONS WILL SUDDENLY STOP AND BEGIN MOVING IN THE OPPOSITE DIRECTION. THIS HAPPENS WHEN THE HEAT IS CONDUCTED TO THE COOL JUNCTION.

SMALL D.C. MOTOR CIRCUITS

USE THESE SIMPLE CIRCUITS TO CONTROL DIRECTION OF ROTATION AND SPEED OF SMALL, LOW-POWER D.C. MOTORS.

MOTOR REVERSERS

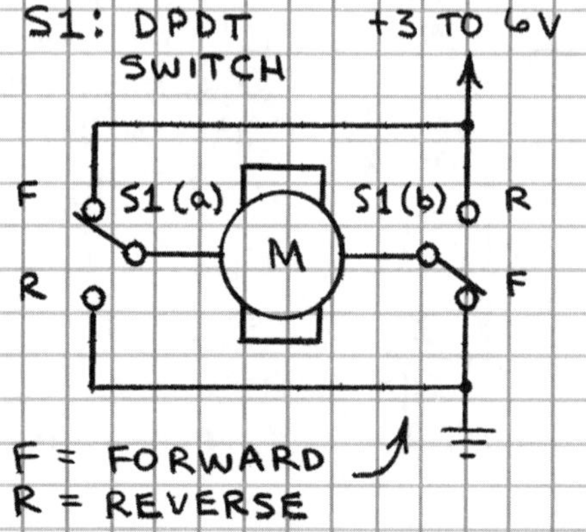

S1: DPDT SWITCH

+3 TO 6V

F S1(a) M S1(b) R

R F

F = FORWARD
R = REVERSE

THESE CIRCUITS CONTROL DIRECTION OF ROTATION OF A MOTOR MANUALLY OR WITH A LOGIC SIGNAL.

IMPORTANT: MOTOR MUST NOT EXCEED POWER RATING OF POWER MOSFETS.

LOW = FORWARD
HIGH = REVERSE

Q1-Q4: IRF-511 OR SIMILAR POWER MOSFET.

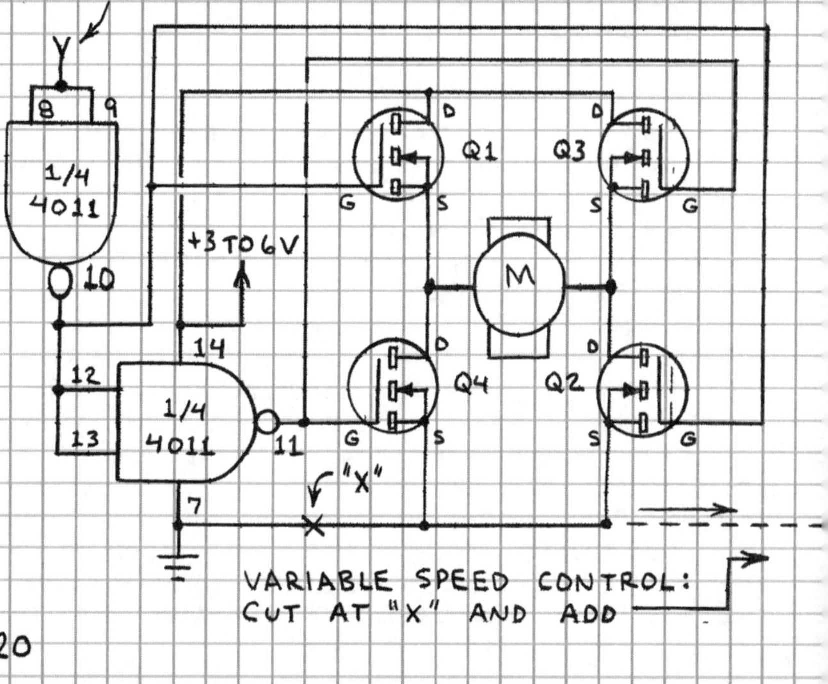

Y 8 9 1/4 4011 10

+3 TO 6V 14 12 13 1/4 4011 11 7

D Q1 G S D Q3 S G M D Q4 G S D Q2 S G

"X"

VARIABLE SPEED CONTROL: CUT AT "X" AND ADD

MOTOR SPEED CONTROLLERS

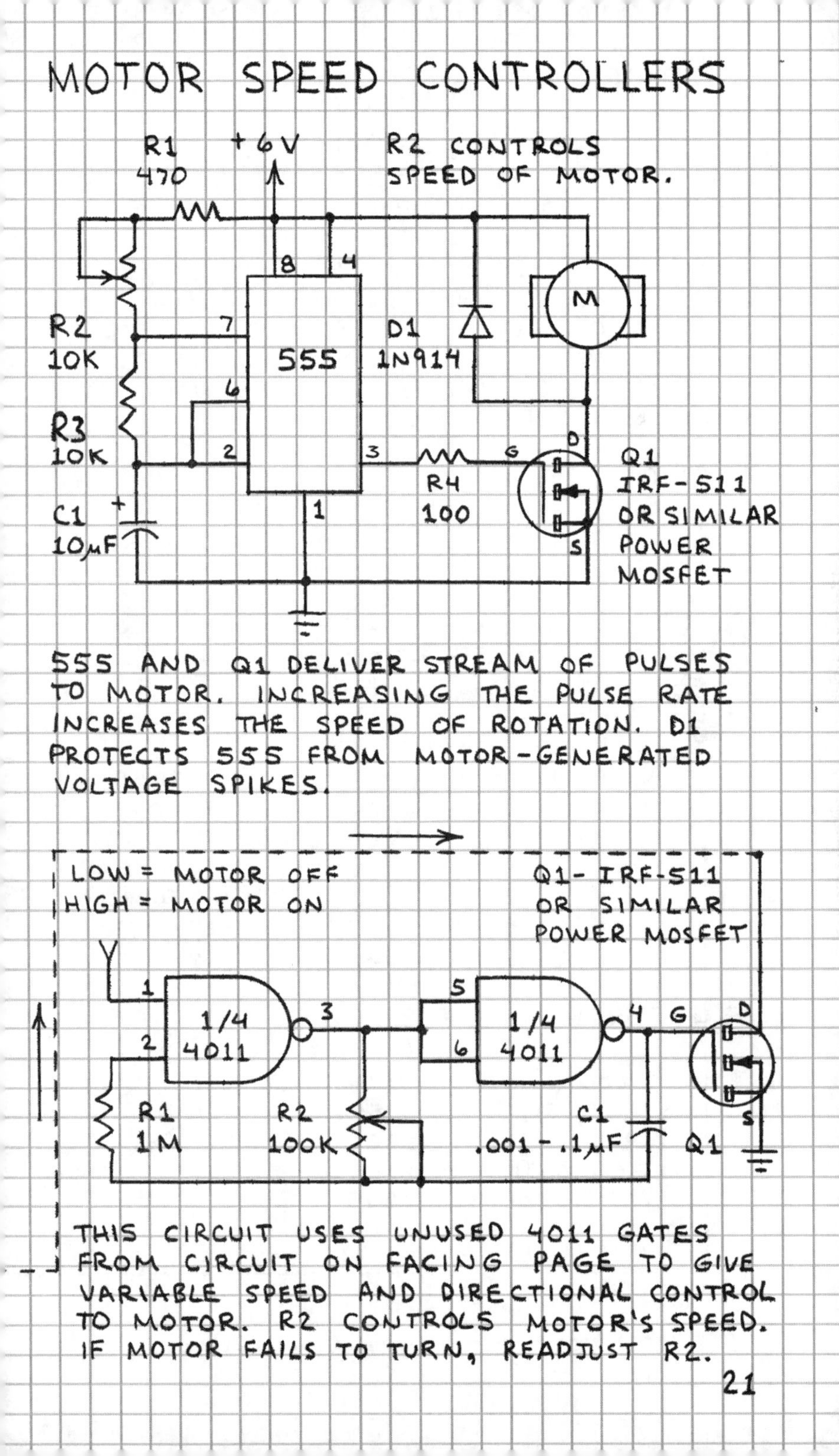

555 AND Q1 DELIVER STREAM OF PULSES TO MOTOR. INCREASING THE PULSE RATE INCREASES THE SPEED OF ROTATION. D1 PROTECTS 555 FROM MOTOR-GENERATED VOLTAGE SPIKES.

THIS CIRCUIT USES UNUSED 4011 GATES FROM CIRCUIT ON FACING PAGE TO GIVE VARIABLE SPEED AND DIRECTIONAL CONTROL TO MOTOR. R2 CONTROLS MOTOR'S SPEED. IF MOTOR FAILS TO TURN, READJUST R2.

INVERSE SQUARE LAW

SOUND WAVES SPREAD OUTWARD AS THEY TRAVEL AWAY FROM THEIR SOURCE. SO DO ELECTROMAGNETIC WAVES SUCH AS LIGHT AND RADIO WAVES. THE INTENSITY OR STRENGTH OF SUCH WAVES IS INVERSELY PROPORTIONAL TO THE SQUARE OF THE DISTANCE OF THE WAVE FROM ITS SOURCE. IN OTHER WORDS, IF THE DISTANCE IS 3, THEN THE INTENSITY IS 1/9 THE INTENSITY WHEN THE DISTANCE IS 1.

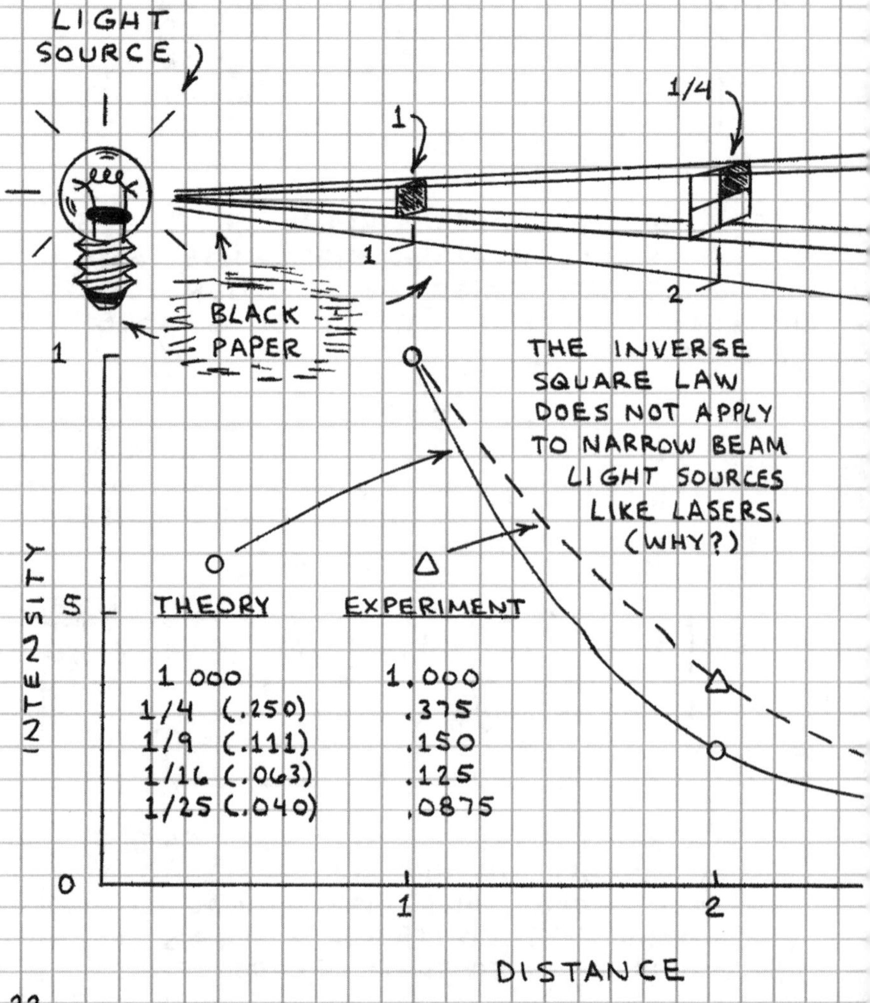

LIGHT SOURCE

1/4

1

1

2

BLACK PAPER

THE INVERSE SQUARE LAW DOES NOT APPLY TO NARROW BEAM LIGHT SOURCES LIKE LASERS. (WHY?)

THEORY (○)	EXPERIMENT (△)
1.000	1.000
1/4 (.250)	.375
1/9 (.111)	.150
1/16 (.063)	.125
1/25 (.040)	.0875

INTENSITY

1

5

0

1

2

DISTANCE

YOU CAN TEST THE INVERSE SQUARE LAW WITH THE HELP OF A SILICON SOLAR CELL AND A STANDARD MULTIMETER SET TO MEASURE CURRENT.

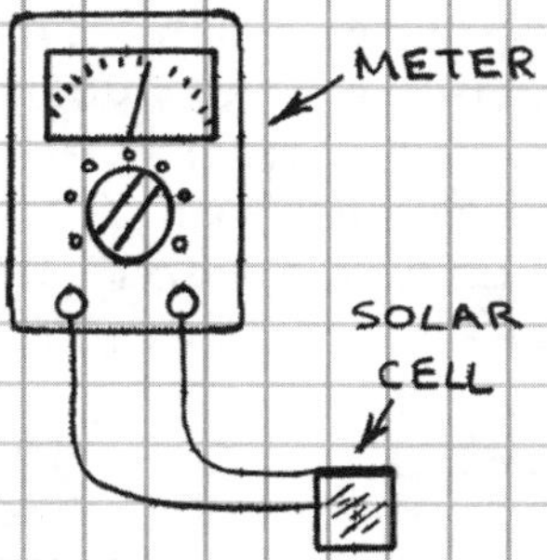

METER

SOLAR CELL

DO THIS EXPERIMENT WITH SUBDUED BACKGROUND LIGHT. PUT LIGHT SOURCE AND SOLAR CELL ON BLACK PAPER.

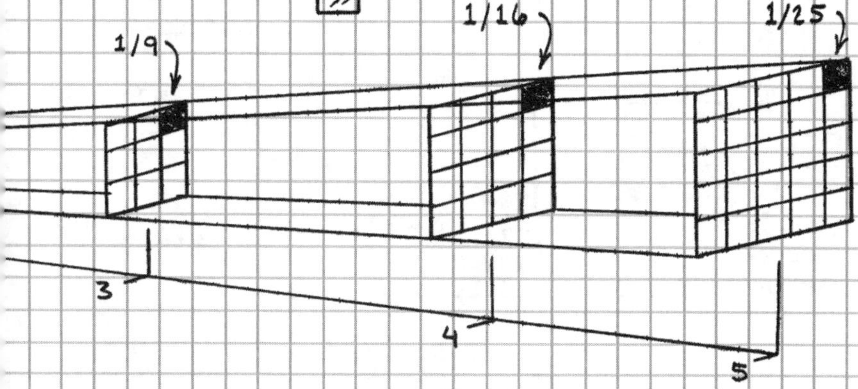

1/9

1/16

1/25

3

4

5

WHY DOES THE EXPERIMENTAL CURVE DIFFER SOMEWHAT FROM THE THEORETICAL CURVE? THE INVERSE SQUARE LAW ASSUMES THAT THE LIGHT SOURCE EMITS UNIFORMLY IN ALL DIRECTIONS. REAL LIGHT SOURCES DO NOT NECESSARILY OBEY THIS ASSUMPTION. FOR BEST RESULTS, THE DISTANCE TO THE FIRST MEASUREMENT POINT SHOULD BE AT LEAST 10 TO 20 TIMES THE SIZE OF THE SOURCE.

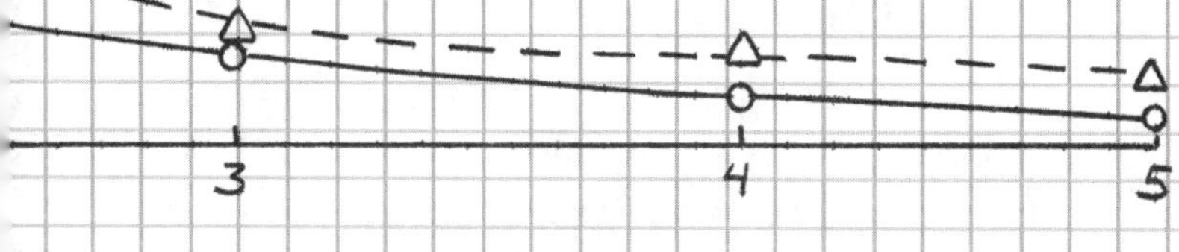

3

4

5

LIGHT LISTENER

THE HUMAN EYE HAS A PERSISTANCE OF VISION OF ABOUT 0.02 SECOND. THEREFORE A LIGHT THAT FLASHES ON AND OFF MORE THAN ABOUT 50 Hz APPEARS CONTINUOUSLY ON THE HUMAN EAR IS MUCH FASTER AND CAN RESPOND TO SOUND WITH A FREQUENCY UP TO ABOUT 20,000 Hz. THE LIGHT LISTENER TRANSFORMS THE PULSATING AND FLICKERING OF LIGHT THAT THE EYE CANNOT DISCERN INTO SOUNDS THE EAR CAN EASILY HEAR.

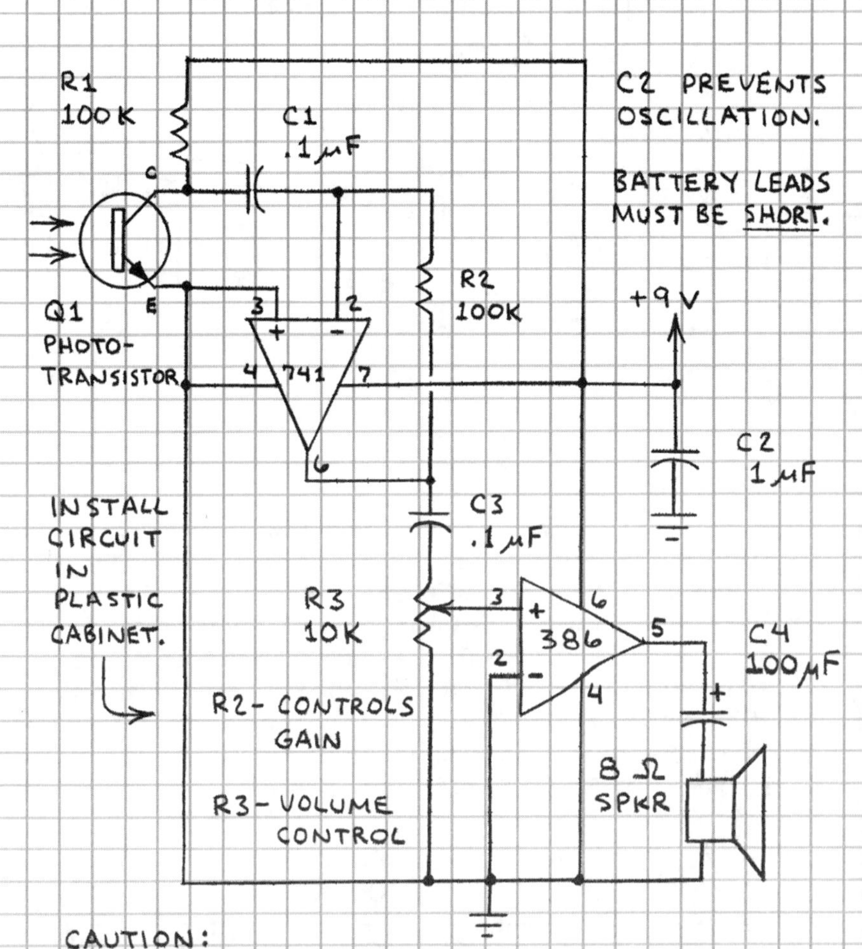

CAUTION:
THIS CIRCUIT CAN PRODUCE LOUD SOUNDS. DO NOT PLACE SPEAKER CLOSE TO YOUR EARS!

TEST THE LIGHT LISTENER BY POINTING Q1 TOWARD AN ARTIFICIAL LIGHT SOURCE. A LINE-POWERED INCANDESCENT LAMP WILL PRODUCE A HUMMING SOUND. A FLUORESCENT LAMP WILL PRODUCE A LOUD BUZZ. AN INFRARED TV REMOTE CONTROL UNIT WILL PRODUCE A PULSING TONE. A CAMERA FLASH UNIT WILL PRODUCE A POP.

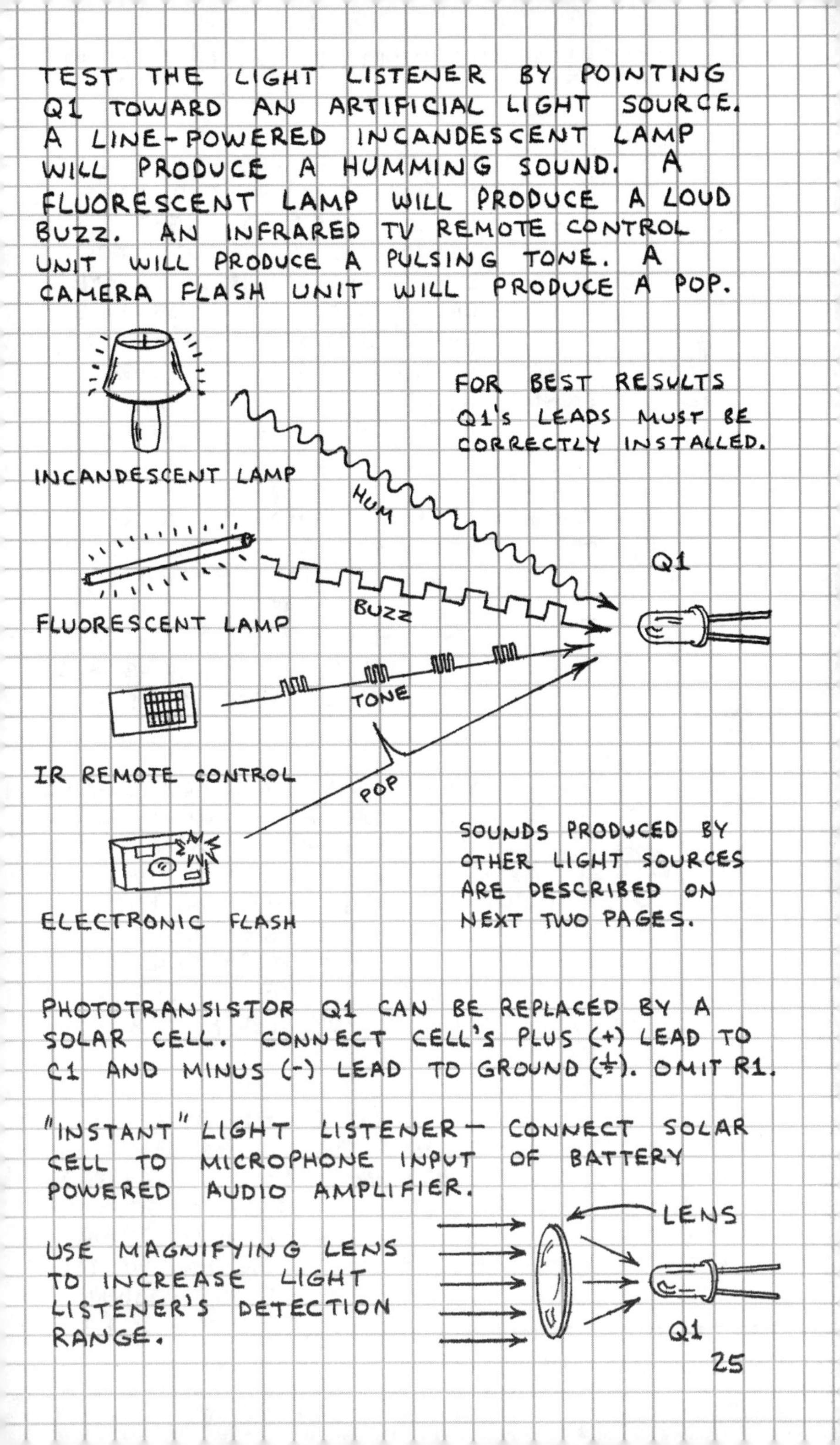

PHOTOTRANSISTOR Q1 CAN BE REPLACED BY A SOLAR CELL. CONNECT CELL'S PLUS (+) LEAD TO C1 AND MINUS (-) LEAD TO GROUND (⏚). OMIT R1.

"INSTANT" LIGHT LISTENER— CONNECT SOLAR CELL TO MICROPHONE INPUT OF BATTERY POWERED AUDIO AMPLIFIER.

USE MAGNIFYING LENS TO INCREASE LIGHT LISTENER'S DETECTION RANGE.

LISTENING TO NATURAL LIGHT

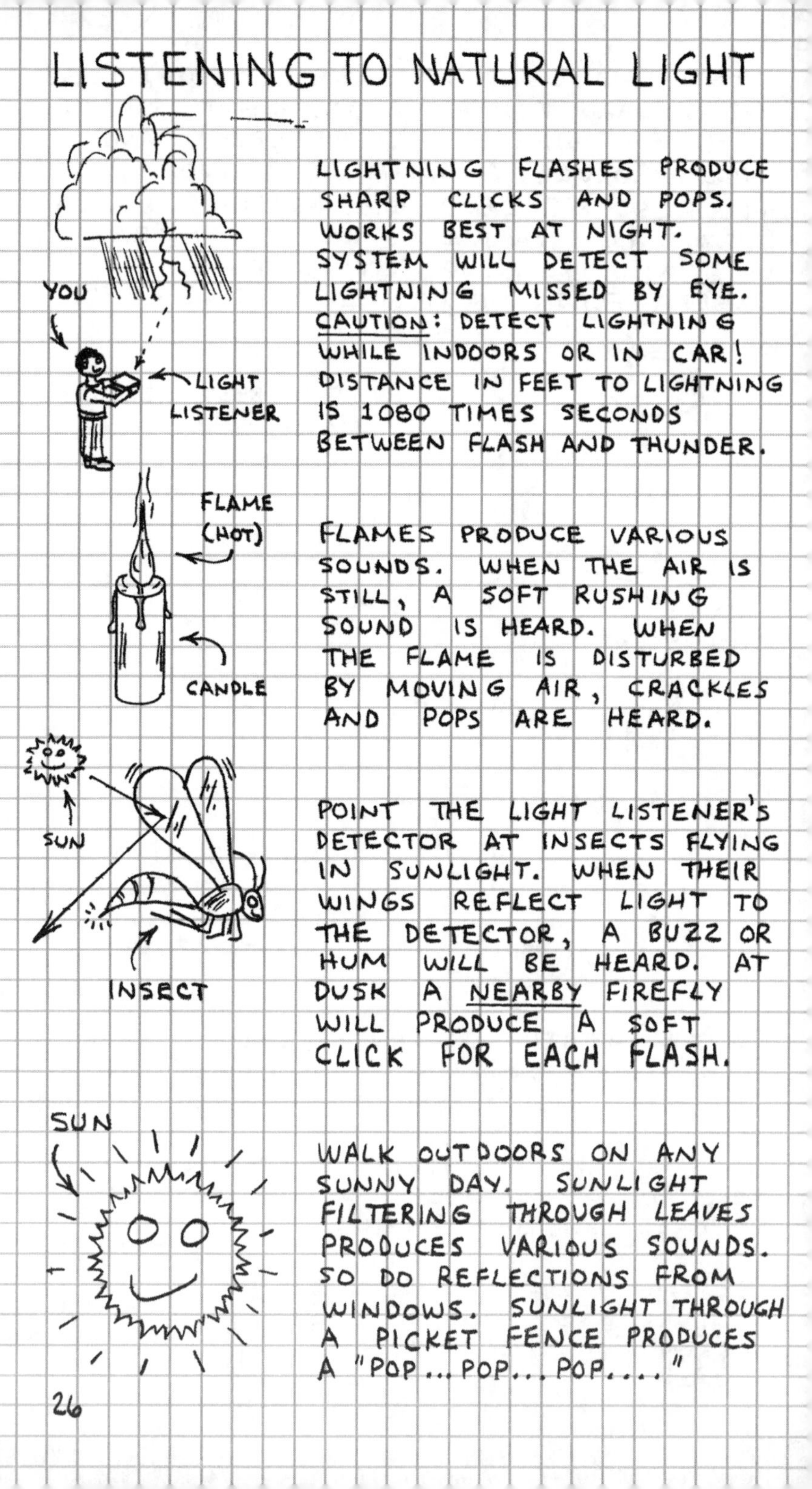

LIGHTNING FLASHES PRODUCE SHARP CLICKS AND POPS. WORKS BEST AT NIGHT. SYSTEM WILL DETECT SOME LIGHTNING MISSED BY EYE. <u>CAUTION</u>: DETECT LIGHTNING WHILE INDOORS OR IN CAR! DISTANCE IN FEET TO LIGHTNING IS 1080 TIMES SECONDS BETWEEN FLASH AND THUNDER.

FLAMES PRODUCE VARIOUS SOUNDS. WHEN THE AIR IS STILL, A SOFT RUSHING SOUND IS HEARD. WHEN THE FLAME IS DISTURBED BY MOVING AIR, CRACKLES AND POPS ARE HEARD.

POINT THE LIGHT LISTENER'S DETECTOR AT INSECTS FLYING IN SUNLIGHT. WHEN THEIR WINGS REFLECT LIGHT TO THE DETECTOR, A BUZZ OR HUM WILL BE HEARD. AT DUSK A <u>NEARBY</u> FIREFLY WILL PRODUCE A SOFT CLICK FOR EACH FLASH.

WALK OUTDOORS ON ANY SUNNY DAY. SUNLIGHT FILTERING THROUGH LEAVES PRODUCES VARIOUS SOUNDS. SO DO REFLECTIONS FROM WINDOWS. SUNLIGHT THROUGH A PICKET FENCE PRODUCES A "POP... POP... POP...."

LISTENING TO ARTIFICIAL LIGHT

SWEEP THE BEAM FROM A FLASHLIGHT ACROSS THE LIGHT LISTENER'S DETECTOR. SLOW SWEEPS PRODUCE A SOFT SWISHING SOUND. FAST SWEEPS GIVE POPS. TAP THE FLASHLIGHT WITH A PENCIL AND A RINGING SOUND WILL BE HEARD AS THE FILAMENT VIBRATES.

BEAM

FLASHLIGHT

THE HEADLIGHTS OF CARS, TRUCKS AND MOTORCYCLES WILL PRODUCE A DISTINCTIVE RINGING SOUND WHEN THE VEHICLE IS MOVING ON A ROUGH OR BUMPY ROAD.

"SINGING" HEADLIGHT

BUMPY ROAD

ELECTRONIC DISPLAYS ARE USUALLY POWERED BY RAPID PULSES OF CURRENT. THE FLASHES ARE MERGED INTO CONTINUOUS LIGHT BY THE SLOW RESPONSE OF THE EYE. BUT THEY CAN BE HEARD AS A BUZZ OR HUM WITH A LIGHT LISTENER.

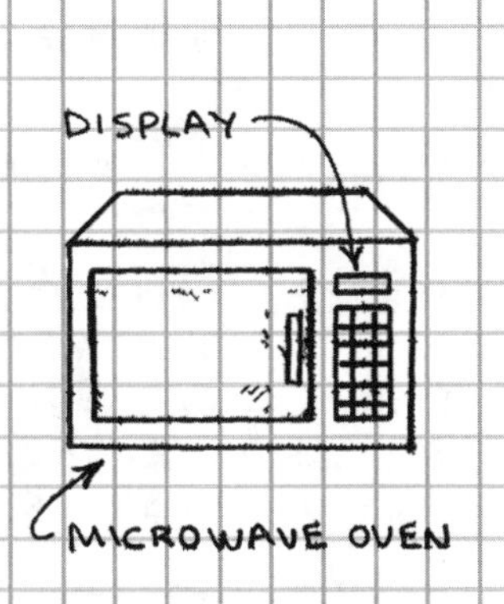

THE DISPLAYS OF TELEVISION SETS AND COMPUTER MONITORS ARE FORMED BY SWEEPING AN ELECTRON BEAM ACROSS A PHOSPHOR COATED SCREEN. THE LIGHT LISTENER TRANSFORMS THE PULSATING PHOSPHOR TO A BUZZ.

TV SET

MONITORING SUNLIGHT

MUCH CAN BE LEARNED ABOUT THE EARTH'S ATMOSPHERE BY MONITORING SUNLIGHT.

THE SOLAR SPECTRUM

CERTAIN GASES ABSORB SPECIFIC WAVELENGTHS OF SUNLIGHT

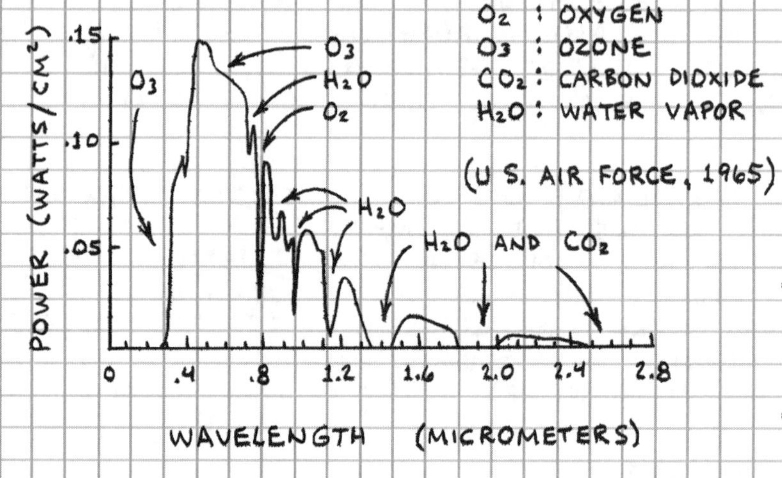

THE SOLAR DAY

THE SOLAR POWER AT THE EARTH'S SURFACE IS INFLUENCED BY THE ATMOSPHERE (CLOUDS, DUST, SMOG, ETC.) AND THE SUN'S ANGLE (TIME OF DAY AND SEASON). HERE'S THE SOLAR POWER FOR A CLEAR SUMMER DAY IN CENTRAL TEXAS:

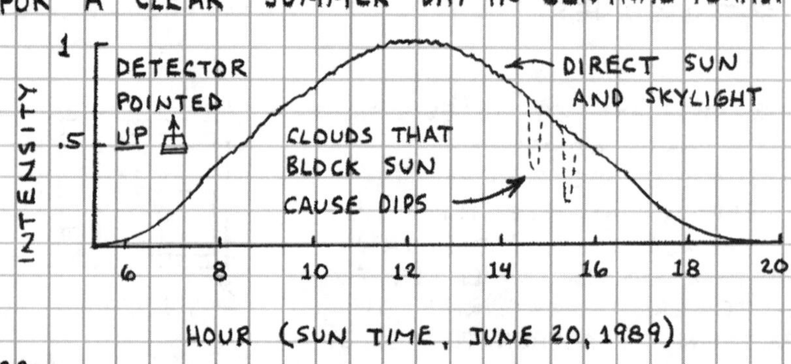

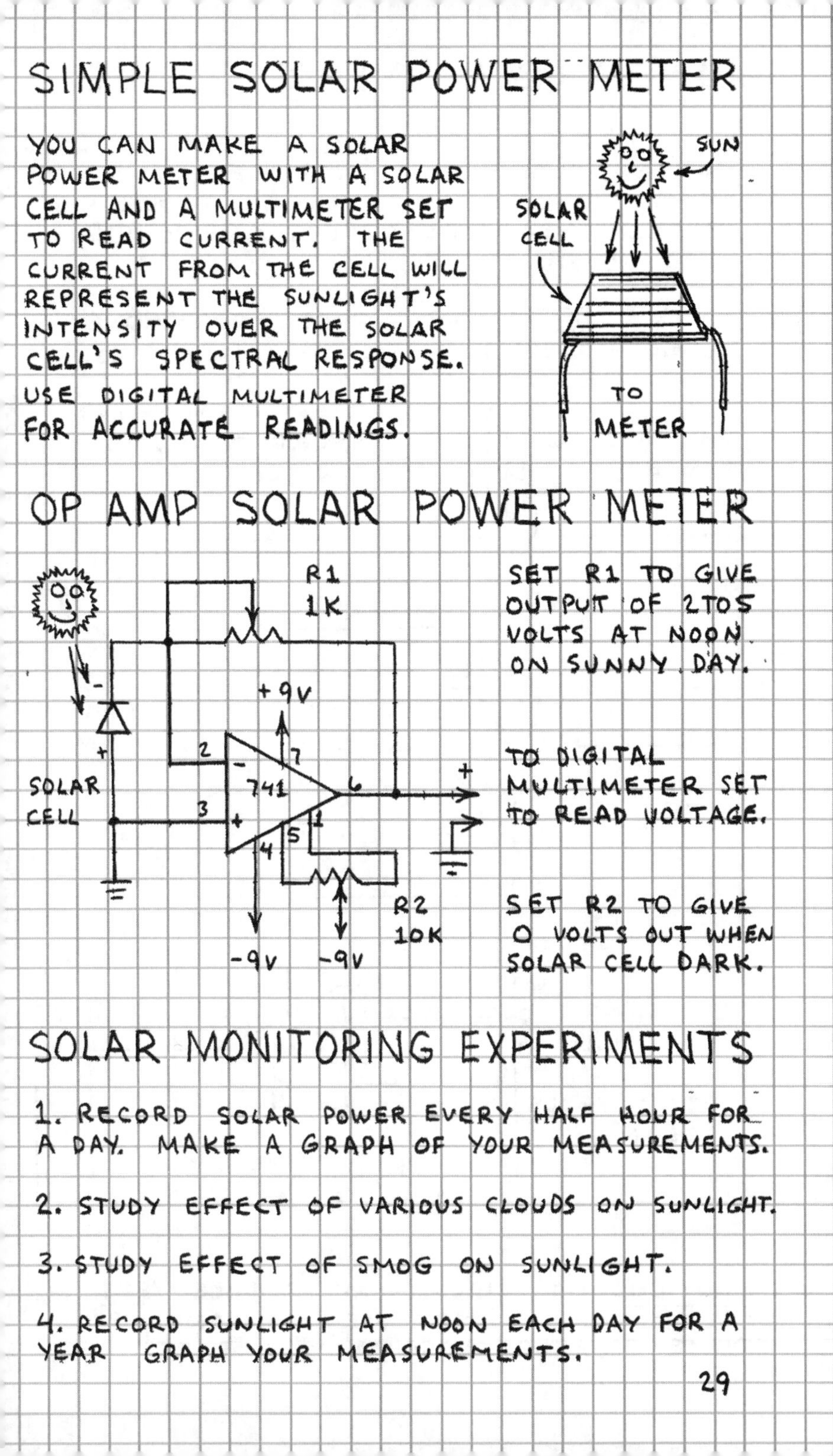

SIMPLE SOLAR POWER METER

YOU CAN MAKE A SOLAR POWER METER WITH A SOLAR CELL AND A MULTIMETER SET TO READ CURRENT. THE CURRENT FROM THE CELL WILL REPRESENT THE SUNLIGHT'S INTENSITY OVER THE SOLAR CELL'S SPECTRAL RESPONSE. USE DIGITAL MULTIMETER FOR ACCURATE READINGS.

OP AMP SOLAR POWER METER

SET R1 TO GIVE OUTPUT OF 2 TO 5 VOLTS AT NOON ON SUNNY DAY.

TO DIGITAL MULTIMETER SET TO READ VOLTAGE.

SET R2 TO GIVE 0 VOLTS OUT WHEN SOLAR CELL DARK.

SOLAR MONITORING EXPERIMENTS

1. RECORD SOLAR POWER EVERY HALF HOUR FOR A DAY. MAKE A GRAPH OF YOUR MEASUREMENTS.

2. STUDY EFFECT OF VARIOUS CLOUDS ON SUNLIGHT.

3. STUDY EFFECT OF SMOG ON SUNLIGHT.

4. RECORD SUNLIGHT AT NOON EACH DAY FOR A YEAR. GRAPH YOUR MEASUREMENTS.

ELECTROMAGNETIC PROBE

ELECTRONIC CIRCUITS THAT OSCILLATE OR SWITCH CURRENT CREATE ELECTROMAGNETIC FIELDS. THIS CIRCUIT CHANGES A PULSING OR OSCILLATING ELECTROMAGNETIC FIELD INTO SOUND.

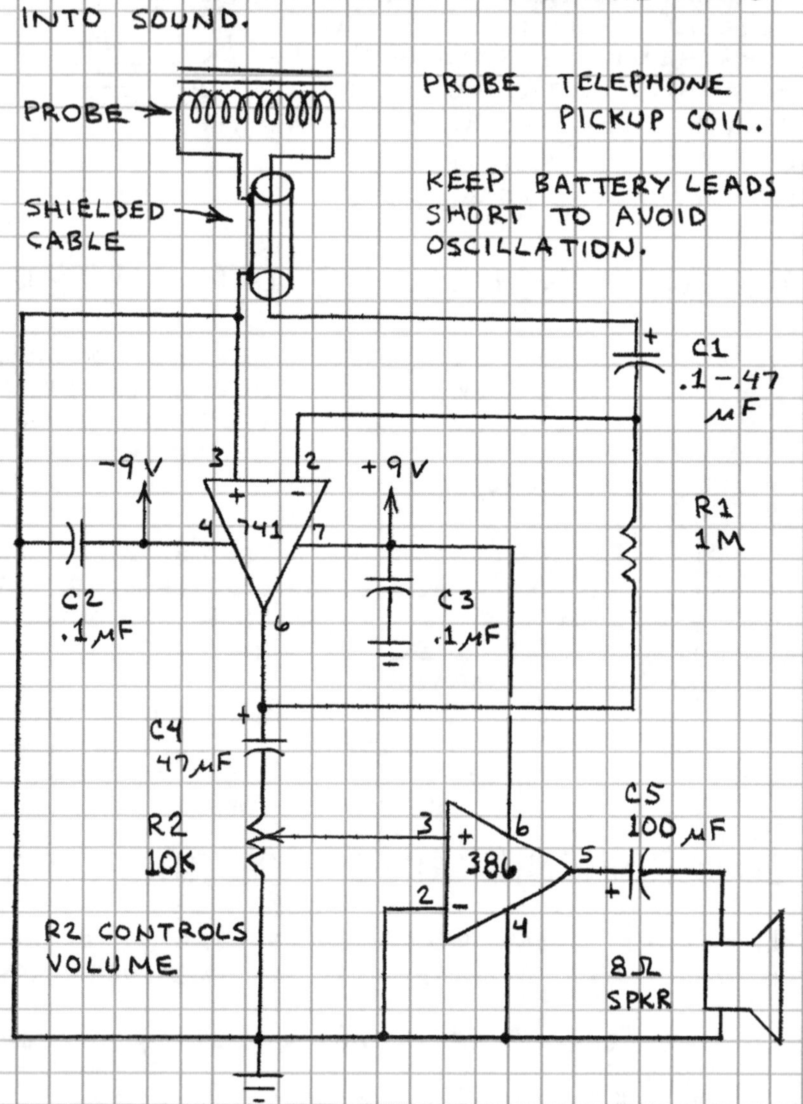

CAUTION: THIS CIRCUIT CAN PRODUCE VERY LOUD SOUNDS. DO NOT USE EARPHONE OR PLACE SPEAKER CLOSE TO YOUR EARS!

USING THE PROBE

TEST THE PROBE BY PLACING PICKUP COIL NEAR RECEIVER OF TELEPHONE HANDSET. YOU SHOULD HEAR A DIAL TONE WHEN HANDSET IS "OFF THE HOOK."

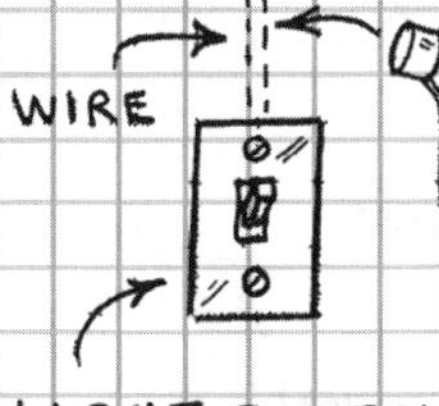

USE PROBE TO FIND WIRES CARRYING ALTERNATING CURRENT YOU CAN FIND WIRES INSIDE WALLS WHEN CURRENT IS FLOWING. TURN SWITCH ON AND HEAR A "POP."

RUB MAGNET AGAINST PICKUP COIL. YOU WILL HEAR RUSHING SOUNDS. IF AMPLIFIER SQUEALS, REDUCE VOLUME (R2). YOU CAN ALSO REDUCE GAIN OF 741 BY REDUCING RESISTANCE OF R1.

NEARBY LIGHTNING FLASHES WILL PRODUCE CRACKLES AND POPS. SPARKS AT BRUSHES OF DIRECT CURRENT MOTORS WILL PRODUCE A BUZZ OR WHINE.

MANY ELECTRONIC APPLIANCES GENERATE ELECTROMAGNETIC FIELDS. TRY PLACING THE PICKUP COIL NEAR COMPUTERS, RADIOS, TELEVISION SETS, FLUORESCENT LIGHTS, RADIO CONTROL TRANSMITTERS, AND INFRARED REMOTE CONTROLLERS.

WIND SPEED INDICATOR

A SMALL D.C. MOTOR WILL GENERATE A VOLTAGE WHEN ITS ARMATURE IS SPUN. THIS PRINCIPLE CAN BE USED TO MAKE A SIMPLE WIND SPEED INDICATOR. THE MOST DIFFICULT ASPECT OF MAKING SUCH AN INSTRUMENT IS MOUNTING AIR COLLECTION CUPS TO THE MOTOR'S SHAFT. THE BEST METHOD IS TO WELD THE CUP HOLDER TO THE SHAFT. HERE IS ONE WAY TO ATTACH AIR CUPS TO A MOTOR FOR TEMPORARY USE:

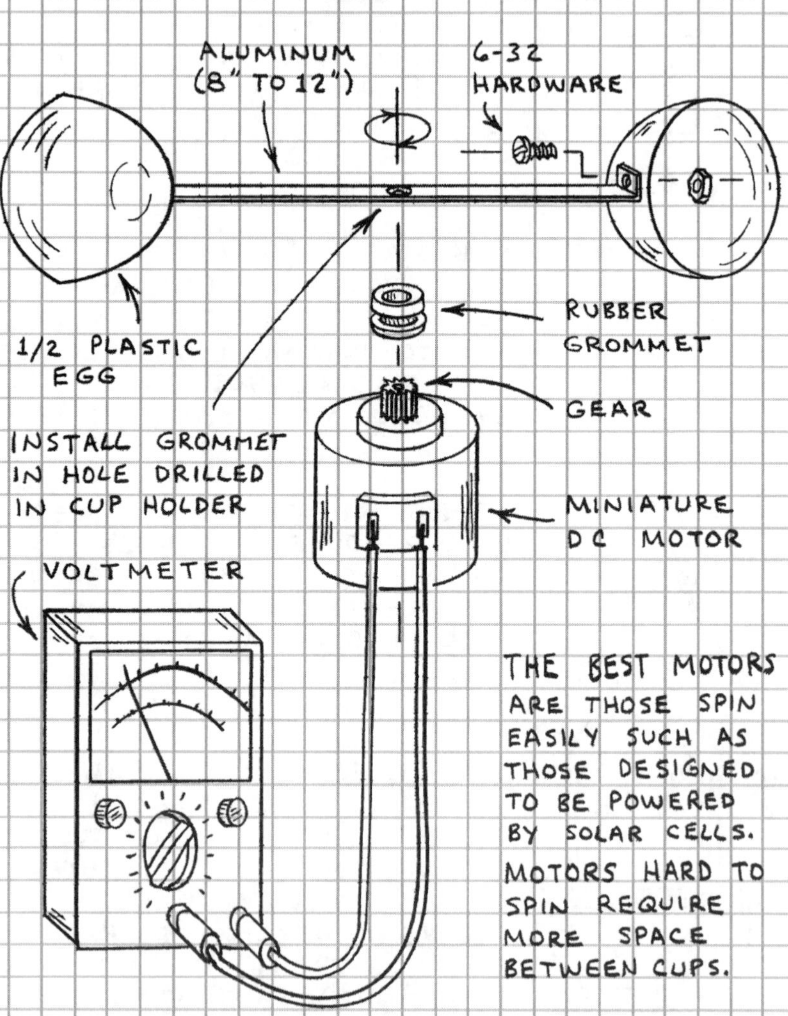

THE BEST MOTORS ARE THOSE SPIN EASILY SUCH AS THOSE DESIGNED TO BE POWERED BY SOLAR CELLS. MOTORS HARD TO SPIN REQUIRE MORE SPACE BETWEEN CUPS.

CALIBRATE THE WIND SPEED INDICATOR WITH A COMMERCIAL ANEMOMETER OR HAVE A FRIEND DRIVE YOU DOWN A COUNTRY ROAD WHILE YOU HOLD THE MAST-MOUNTED UNIT (SEE BELOW) OUT A PASSENGER-SIDE WINDOW. RECORD THE MOTOR'S VOLTAGE AT VARIOUS SPEEDS AND MAKE A CALIBRATION GRAPH LIKE THIS.

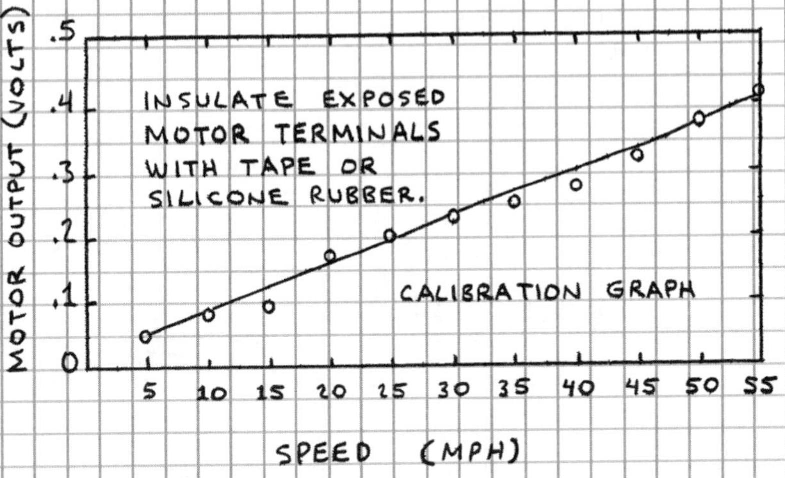

MAST INSTALLATION

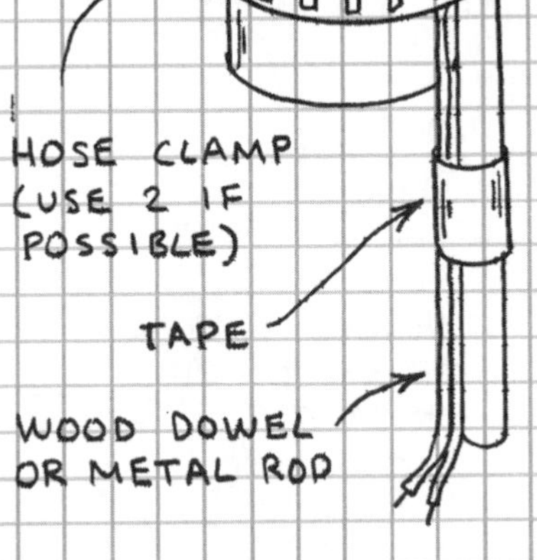

CAUTION:

1 NEVER HOLD THE UNIT AT EYE LEVEL WHEN THE CUPS ARE SPINNING!

2. DO NOT INSTALL THE UNIT NEAR A POWER LINE!

3. USE GREAT CARE WHEN CALIBRATING THE UNIT FROM A MOVING CAR!

RAIN SENSORS

RAIN DROPS CONDUCT ELECTRICITY. THIS MEANS THAT A SIMPLE RAIN DETECTOR CAN BE MADE FROM TWO CLOSELY SPACED ELECTRODES. THE CHANCE OF DETECTING A SINGLE RAIN DROP ARE INCREASED BY INCREASING THE AREA OF THE ELECTRODES. HERE ARE SEVERAL WAYS TO MAKE RAIN SENSORS.

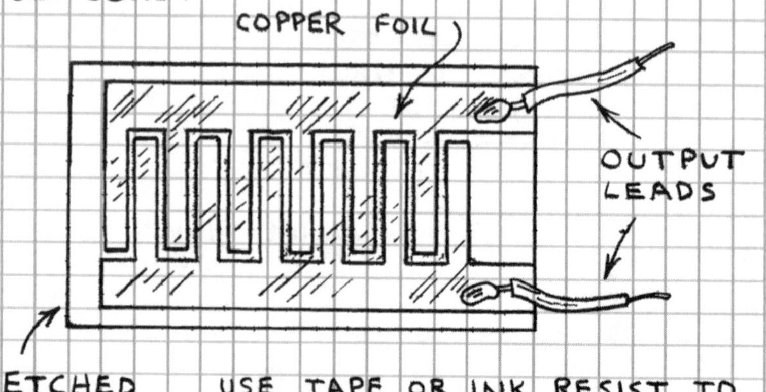

ETCHED CIRCUIT BOARD

USE TAPE OR INK RESIST TO MAKE ELECTRODE PATTERN. THEN ETCH. REMOVE RESIST.

<u>NOTE</u>: COPPER MUST BE SHINY BRIGHT BEFORE SOLDERING!

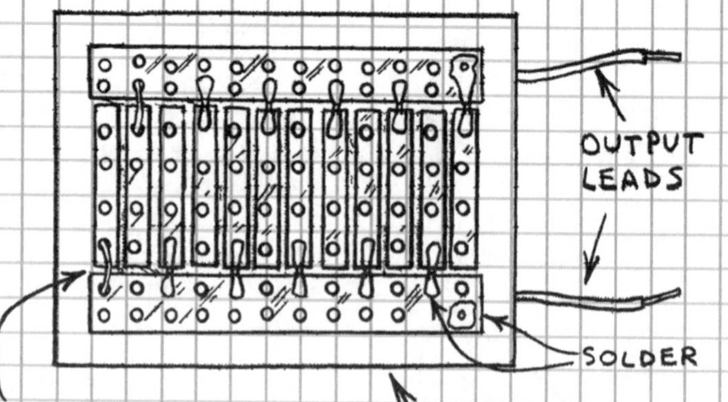

INSERT WIRES BETWEEN ALTERNATING ELECTRODES AND SOLDER IN PLACE.

RADIO SHACK PRE-ETCHED CIRCUIT BOARD SEGMENT.

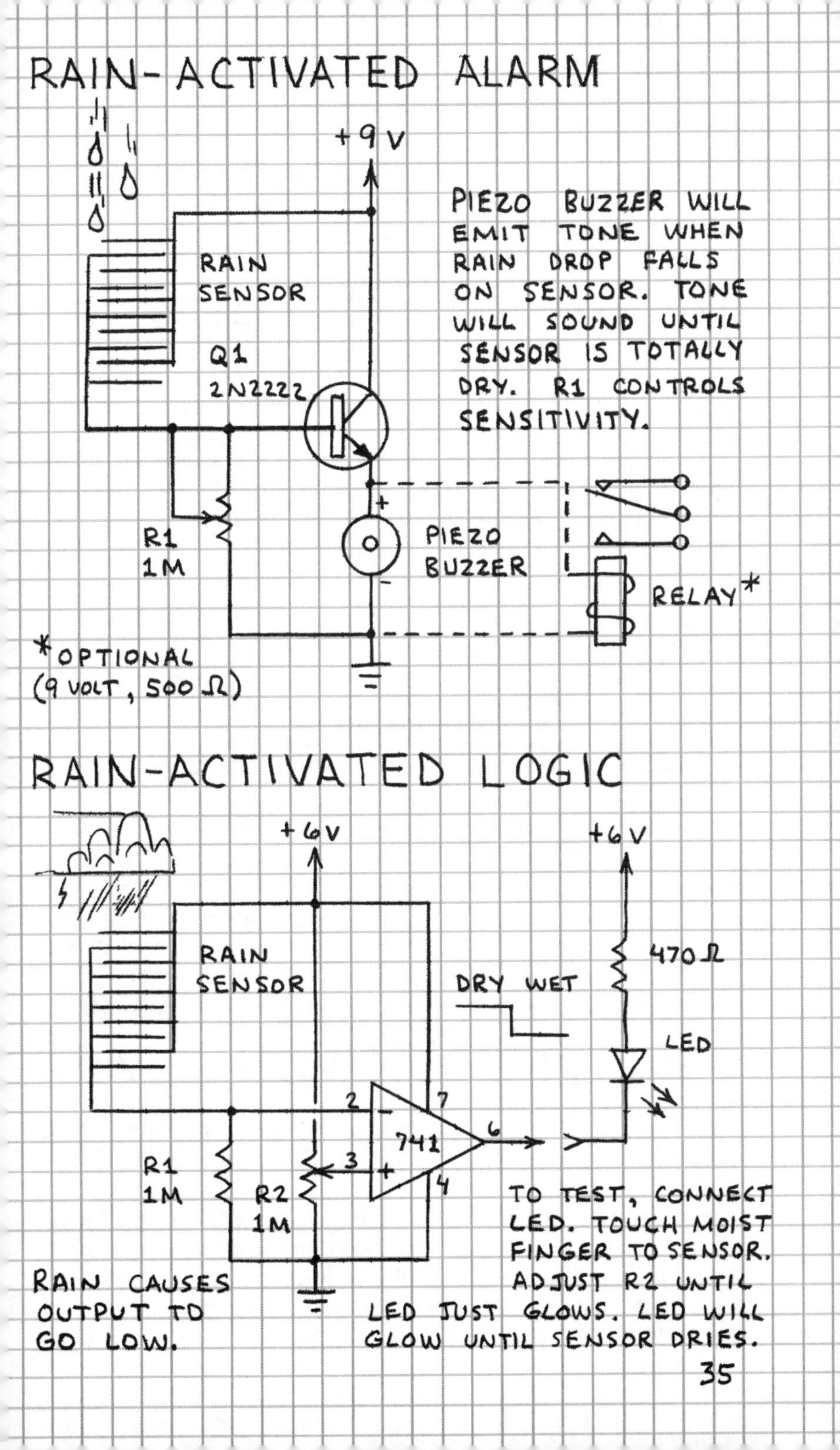
RAIN-ACTIVATED ALARM
+9V
RAIN SENSOR
Q1 2N2222
PIEZO BUZZER WILL EMIT TONE WHEN RAIN DROP FALLS ON SENSOR. TONE WILL SOUND UNTIL SENSOR IS TOTALLY DRY. R1 CONTROLS SENSITIVITY.
R1 1M
+
PIEZO BUZZER
-
RELAY*
*OPTIONAL (9 VOLT, 500 Ω)
RAIN-ACTIVATED LOGIC
+6V
+6V
RAIN SENSOR
DRY WET
470 Ω
LED
2
7
741
6
3
4
R1 1M
R2 1M
TO TEST, CONNECT LED. TOUCH MOIST FINGER TO SENSOR. ADJUST R2 UNTIL LED JUST GLOWS. LED WILL GLOW UNTIL SENSOR DRIES.
RAIN CAUSES OUTPUT TO GO LOW.
35

ELECTRONIC THERMOMETER

A THERMISTOR IS A TEMPERATURE DEPENDENT RESISTOR. THERMISTORS CAN BE USED TO MAKE VARIOUS KINDS OF ELECTRONIC THERMOMETERS.

THERMISTOR CIRCUITS

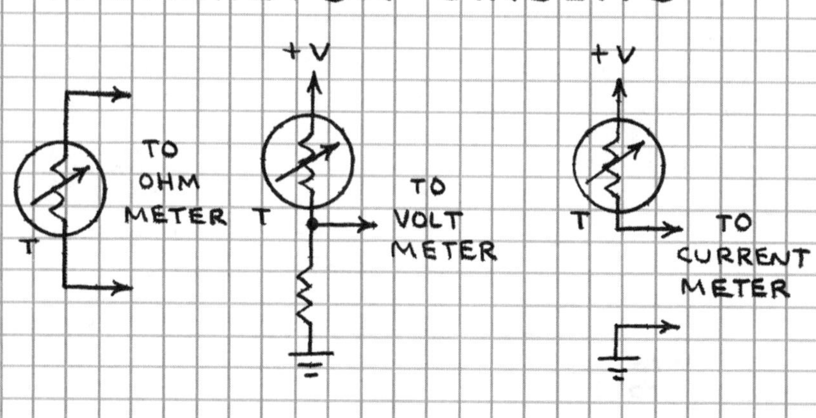

THERMISTOR AMPLIFIER

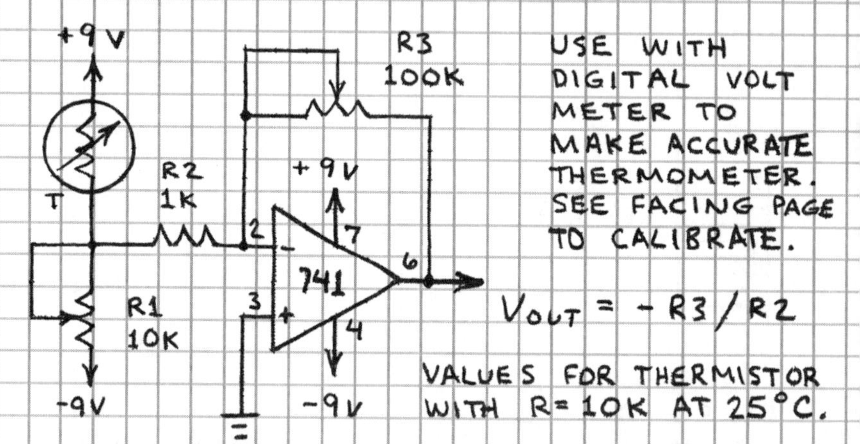

USE WITH DIGITAL VOLT METER TO MAKE ACCURATE THERMOMETER. SEE FACING PAGE TO CALIBRATE.

$V_{OUT} = -R3/R2$

VALUES FOR THERMISTOR WITH R= 10K AT 25°C.

INCREASE R3'S RESISTANCE TO INCREASE SENSITIVITY OVER SMALL TEMPERATURE RANGE. REDUCE R3'S RESISTANCE TO REDUCE SENSITIVITY OVER LARGE TEMPERATURE RANGE. R1 IS ZERO ADJUST.

THERMISTOR CALIBRATION

WATER PROOF THERMISTOR LEADS WITH SILICONE SEALANT AND ALLOW TO CURE. DIP THERMISTOR IN HOT WATER AND RECORD RESISTANCE, VOLTAGE OR CURRENT AS TEMPERATURE OF WATER FALLS. ADD ICE TO SPEED COOLING.

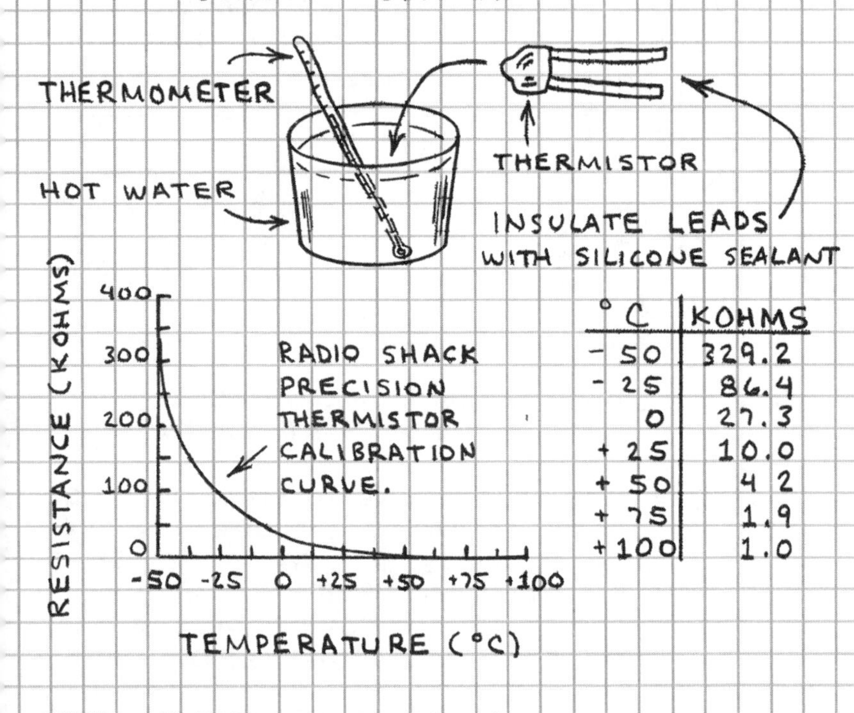

°C	KOHMS
−50	329.2
−25	86.4
0	27.3
+25	10.0
+50	4.2
+75	1.9
+100	1.0

TEMPERATURE SWITCH

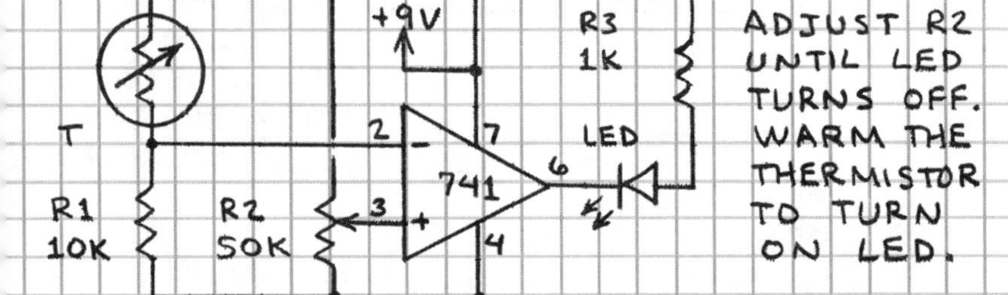

ADJUST R2 UNTIL LED TURNS OFF. WARM THE THERMISTOR TO TURN ON LED.

OK TO REPLACE R3 AND LED WITH RELAY.

REVERSE CONNECTIONS TO PINS 2 AND 3 TO REVERSE OPERATION.

MOTION DETECTOR

WHEN PROPERLY ADJUSTED, THIS SIMPLE CIRCUIT WILL DETECT THE MOVEMENT OF AN OBJECT WITHIN ITS FIELD OF VIEW. THE DETECTION RANGE CAN BE TENS OF FEET.

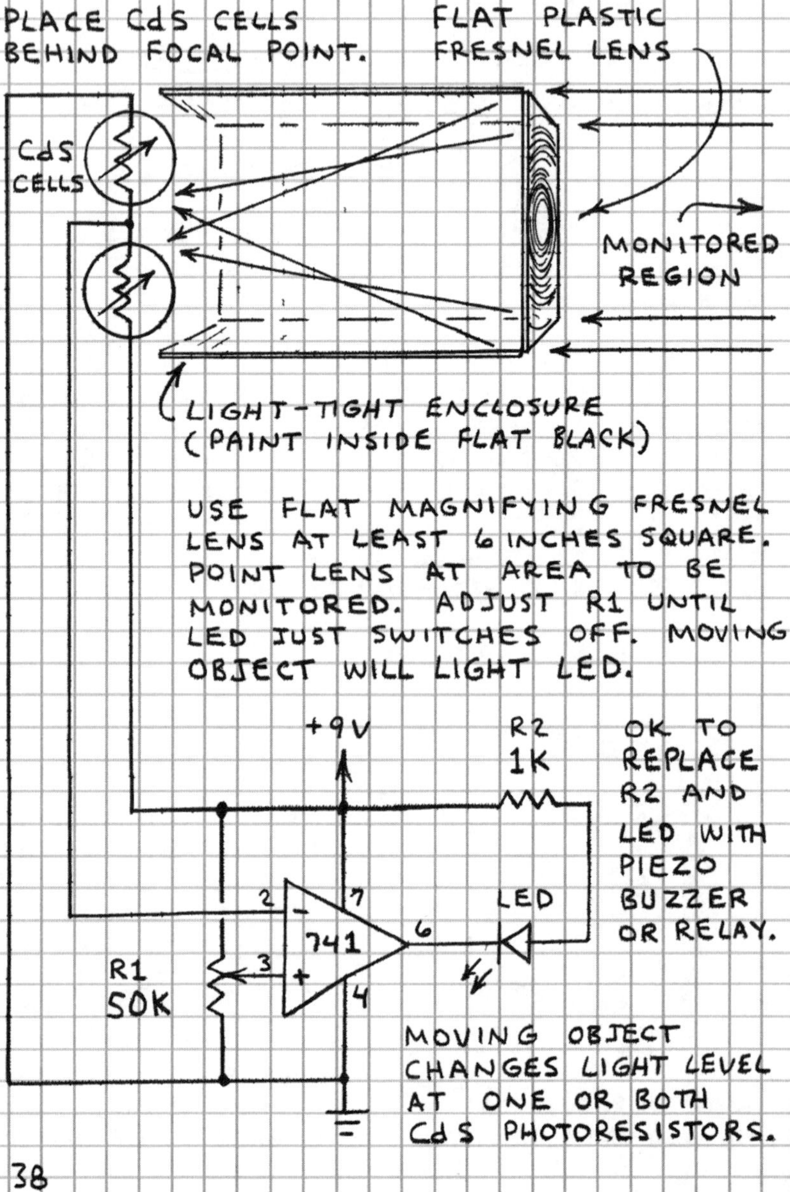

PLACE CdS CELLS BEHIND FOCAL POINT.

FLAT PLASTIC FRESNEL LENS

CdS CELLS

MONITORED REGION

LIGHT-TIGHT ENCLOSURE (PAINT INSIDE FLAT BLACK)

USE FLAT MAGNIFYING FRESNEL LENS AT LEAST 6 INCHES SQUARE. POINT LENS AT AREA TO BE MONITORED. ADJUST R1 UNTIL LED JUST SWITCHES OFF. MOVING OBJECT WILL LIGHT LED.

+9V

R2 1K

OK TO REPLACE R2 AND LED WITH PIEZO BUZZER OR RELAY.

2

7

LED

741

6

3

R1 50K

4

MOVING OBJECT CHANGES LIGHT LEVEL AT ONE OR BOTH CdS PHOTORESISTORS.

POSITION DETECTOR

USE THIS CIRCUIT TO INDICATE POSITION OF A BEAM OF LIGHT FALLING ON TWO ADJACENT SOLAR CELLS.

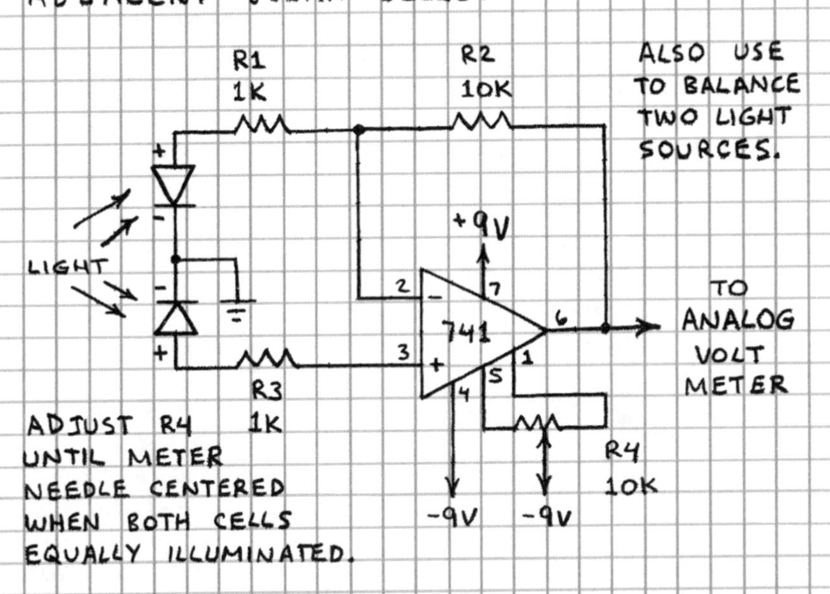

TEST WITH SUPER BRIGHT LED IN CLEAR PACKAGE. BEAM STRUCTURE MAY AFFECT READOUT. IF ONE CELL GENERATES MORE VOLTAGE WHEN LIGHT LEVEL IS BALANCED, REDUCE OTHER CELL'S INPUT RESISTANCE (R1 OR R3).

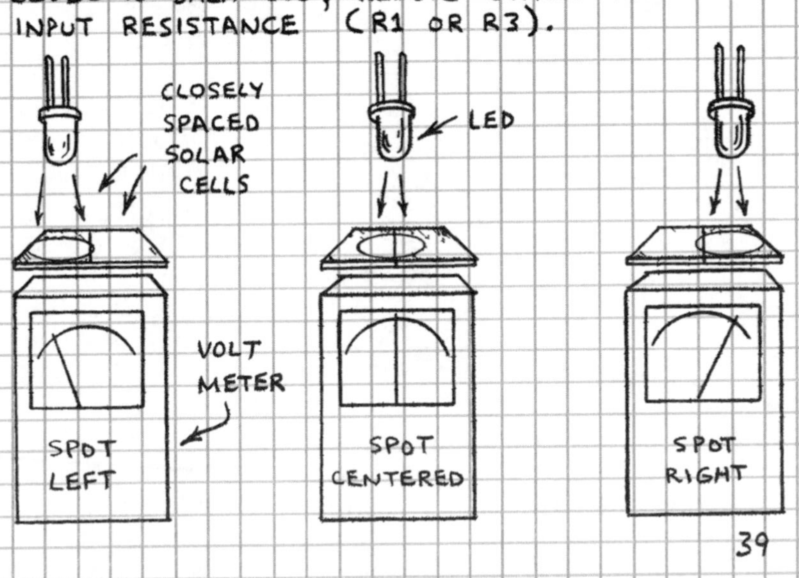

PRESSURE SENSOR

THE CONDUCTIVE FOAM PLASTIC IN WHICH STATIC-SENSITIVE COMPONENT LEADS ARE INSERTED CAN BE USED TO MAKE PRESSURE-SENSITIVE RESISTORS. YOU CAN USE A PAIR OF SUCH RESISTORS TO MAKE A PRESSURE-SENSITIVE COMPUTER JOYSTICK. A PRESSURE-SENSITIVE RESISTOR CAN BE USED TO MAKE AN ELECTRONIC SCALE. A SIMPLE ACCELEROMETER CAN BE MADE BY ATTACHING A LEAD FISHING WEIGHT TO THE MOVABLE CONTACT OF A PRESSURE-SENSITIVE RESISTOR. HERE IS ONE OF MANY WAYS TO MAKE A PRESSURE-SENSITIVE RESISTOR:

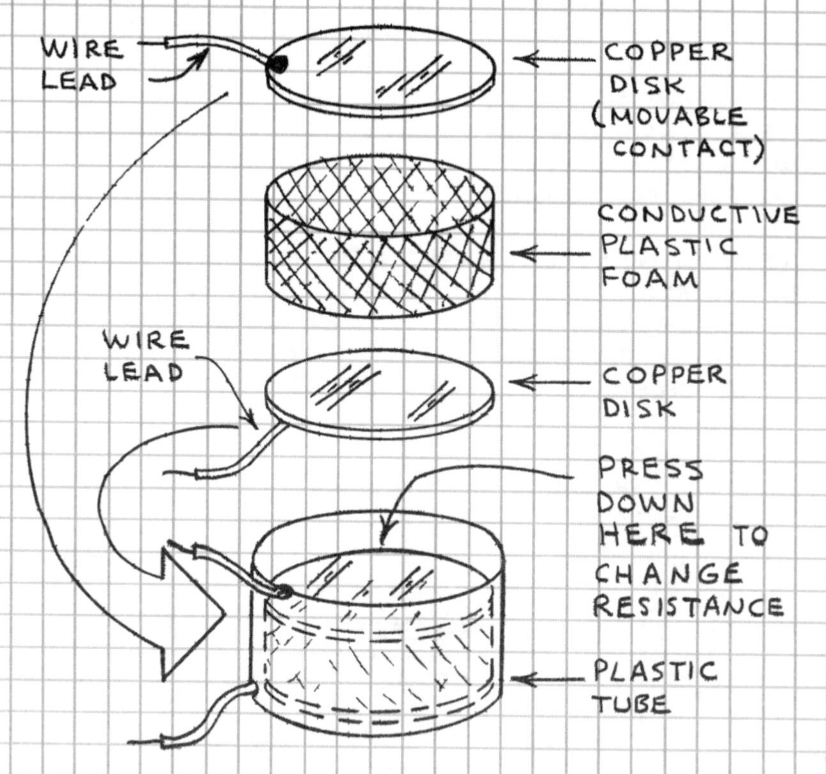

COPPER DISKS CAN BE PENNYS, COPPER FOIL OR COPPER-CLAD CIRCUIT BOARD. POLISH COPPER BEFORE SOLDERING LEADS.

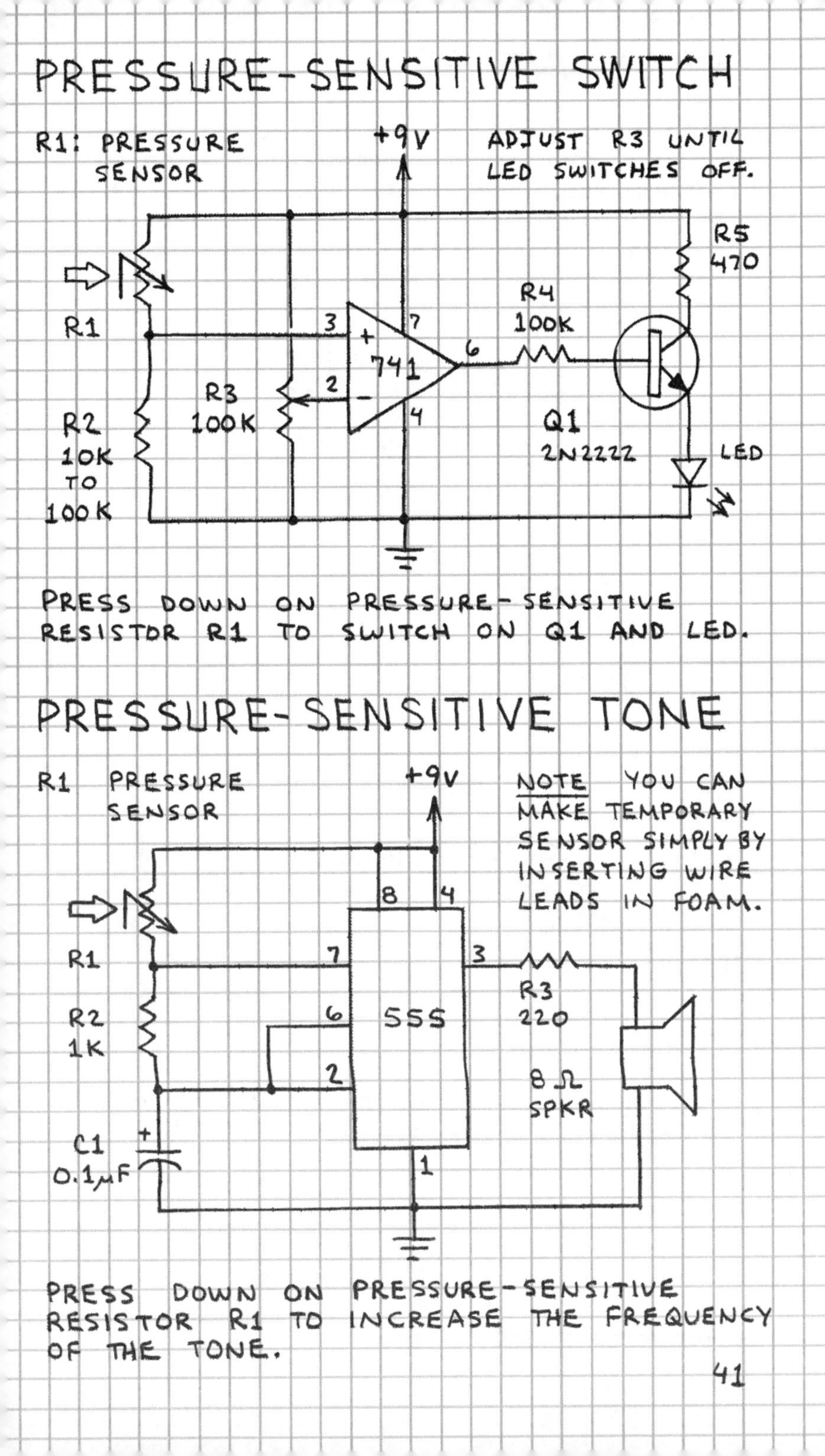
PRESSURE-SENSITIVE SWITCH
R1: PRESSURE SENSOR
+9V
ADJUST R3 UNTIL LED SWITCHES OFF.
R5 470
R4 100K
R1
3
+
7
741
6
2
−
4
R3 100K
R2 10K TO 100K
Q1 2N2222
LED
PRESS DOWN ON PRESSURE-SENSITIVE RESISTOR R1 TO SWITCH ON Q1 AND LED.
PRESSURE-SENSITIVE TONE
R1 PRESSURE SENSOR
+9V
NOTE YOU CAN MAKE TEMPORARY SENSOR SIMPLY BY INSERTING WIRE LEADS IN FOAM.
8
4
R1
7
3
R3 220
R2 1K
6
555
2
8 Ω SPKR
C1 0.1μF
+
1
PRESS DOWN ON PRESSURE-SENSITIVE RESISTOR R1 TO INCREASE THE FREQUENCY OF THE TONE.
41

SEISMOMETER

A SEISMOMETER IS AN INSTRUMENT THAT DETECTS THE EARTH MOVEMENTS CAUSED BY EARTHQUAKES. A SIMPLE SEISMOMETER CAN DETECT EARTHQUAKES THAT OCCUR THOUSANDS OF MILES AWAY. EARTHQUAKES CAUSE SEVERAL KINDS OF SEISMIC WAVES INSIDE THE EARTH.

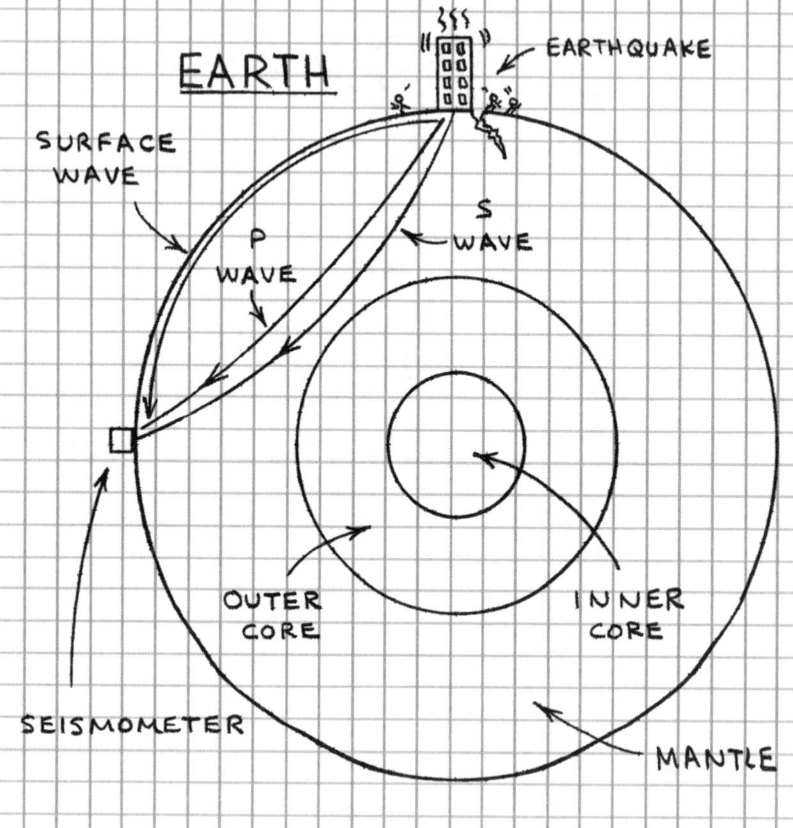

A SEISMOGRAM IS A GRAPH OF THE EARTH'S MOVEMENTS PRODUCED BY A SEISMOMETER.

P WAVE S WAVE SURFACE WAVE

P WAVE ARRIVES FIRST.

MANY DIFFERENT KINDS OF SEISMOMETERS ARE AVAILABLE. TWO EXAMPLES:

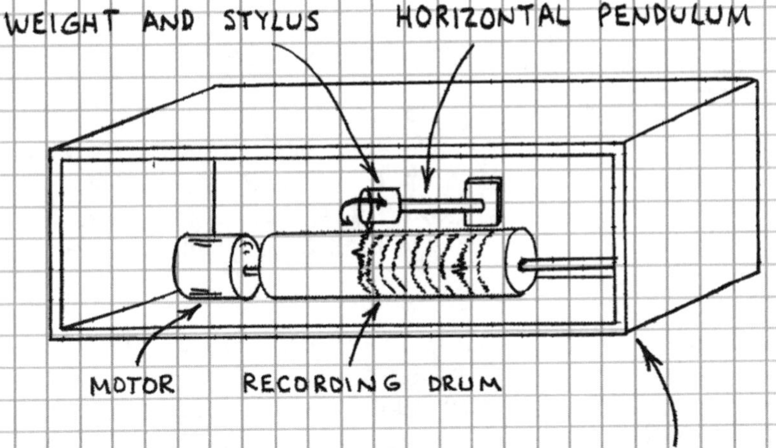

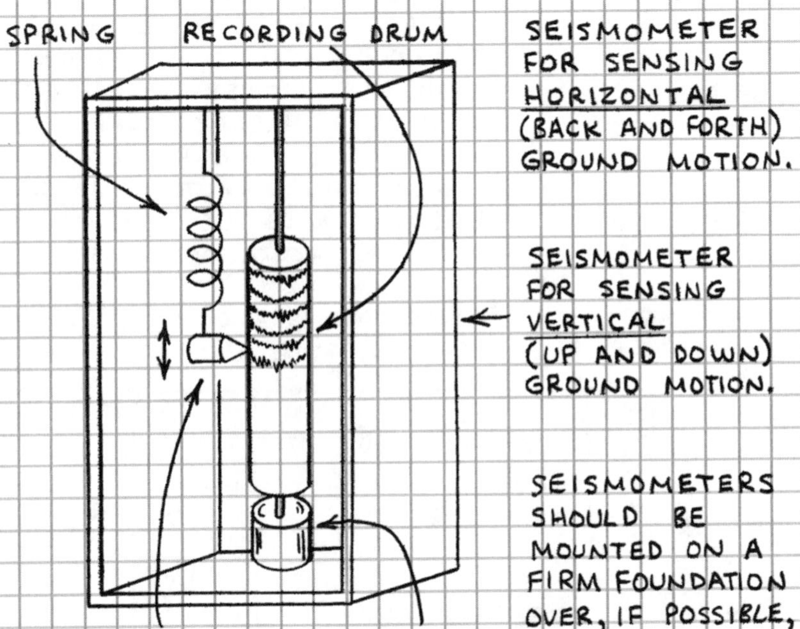

FOR MORE INFORMATION ABOUT SEISMOMETERS, VISIT A LIBRARY. "EARTHQUAKES" (W H. FREEMAN AND CO., 1988) BY BRUCE BOLT IS A GOOD BOOK.

EARTH MOVEMENT SENSOR

THIS SIMPLE SEISMIC SENSOR HAS DETECTED TRAINS MORE THAN ONE MILE AWAY.

STURDY BEAM

USE THIS TEMPORARY PENDULUM FOR INITIAL TESTS. INSTALL PICKUP COIL ON HEAVY SURFACE THAT CAN BE MOVED DIRECTLY UNDER MAGNET.

2 TO 4 FEET WIRE OR NYLON LINE

ADJUSTMENT PROCEDURE. PLACE A MAGNET DIRECTLY ON TOP OF PICKUP COIL. ADJUST R2 UNTIL LED JUST SWITCHES OFF AND DOES NOT FLICKER. REMOVE MAGNET. LED SHOULD FLASH AND FLICKER WHEN THE MAGNET IS MOVED NEAR THE PICKUP COIL. NEXT, PLACE PICKUP COIL DIRECTLY UNDER PENDULUM MAGNET. REDUCE R1 TO 1M IF CIRCUIT TOO SENSITIVE.

AVOID MOVING AIR!

MAGNET

HOT MELT GLUE

REDUCE THIS SPACE TO INCREASE SENSITIVITY.

R1
10M

+9V

USE SHIELDED CABLE IF MORE THAN 2-3" LONG.

2 − 7
741
6
3 + 4

-9V

PICKUP COIL (USE TELEPHONE PICKUP COIL OR 9-VOLT RELAY).

KEEP BATTERY LEADS SHORT.

+9V + − + − -9V

9V 9V

THIS SENSOR IS EXCEPTIONALLY SENSITIVE! IF LED FAILS TO STOP FLICKERING, REDUCE THE SENSITIVITY. EITHER READJUST R2 OR INCREASE SPACE BETWEEN MAGNET AND PICKUP COIL.

FOR PERMANENT USE INSTALL PENDULUM AND PICKUP COIL IN METAL OR PLASTIC PIPE TO PREVENT AIR MOVEMENT FROM MOVING PENDULUM. USE L-BRACKETS TO BOLT ASSEMBLY TO CONCRETE FOUNDATION FOR BEST RESULTS. WHEN HE WAS IN HIGH SCHOOL IN TEXAS, ERIC RYAN MIMS USED SIMILAR ARRANGEMENT TO DETECT UNDERGROUND NUCLEAR TESTS IN NEVADA.

ADJUST POSITION OF MAGNET BY MOVING CAP UP OR DOWN OR BY MOVING PENDULUM THROUGH HOLE IN CAP. MAKE OBSERVATION PORT TO OBSERVE MAGNET. COVER WITH CLEAR PLASTIC WINDOW.

THREADED OR PUSH-ON CAP

OBSERVATION PORT

PICKUP COIL

+9V

3 + 741 − 2

7 6 4

LED R3 470 +9V

+9V

R2 10K

−9V

OPTIONAL PIEZO BUZZER; CHIRPS WHEN LED FLASHES.

OK TO REPLACE LED WITH BUZZER OR TO USE BOTH. INCREASE R3 TO REDUCE VOLUME.

RF TELEMETRY TRANSMITTER

THIS SIMPLE LOW-POWER RADIO FREQUENCY (RF) TRANSMITTER WILL BROADCAST TEMPERATURE AS A SERIES OF CLICKS TO A NEARBY RADIO TUNED TO THE UPPER END OF THE AM BROADCAST BAND.

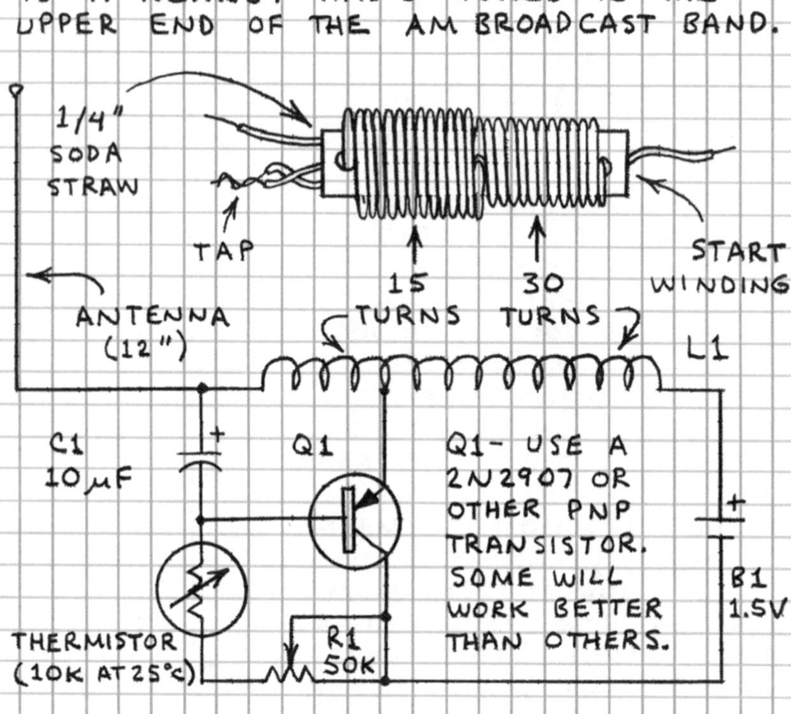

L1 USE 30 GAUGE WRAPPING WIRE OR MAGNET WIRE. (USE MAGNET WIRE FOR SMALLER COIL. BURN VARNISH FROM ENDS OF WIRE AND TAP AND LIGHTLY BUFF CHARRED VARNISH WITH SAND PAPER.) PUNCH SMALL HOLE NEAR ONE END OF STRAW. INSERT 2" OF WIRE THROUGH HOLE AND WIND 30 TURNS. PUNCH HOLE IN STRAW AND INSERT 2" LOOP OF WIRE (TAP) THROUGH HOLE. WIND 15 TURNS BACK OVER FIRST WINDING. PUNCH HOLE THROUGH WINDING AND INSERT END OF WIRE. WRAPPING WIRE: CUT TAP LOOP AND TWIST EXPOSED WIRES.

C1: INCREASE VALUE TO SLOW PULSE RATE.

R1: ADJUST TO CHANGE PULSE RATE.

B1: USE AA PENLIGHT CELL.

SAMPLE CALIBRATION GRAPH

WATERPROOF LEADS OF THERMISTOR WITH SILICONE SEALANT. IMMERSE THERMISTOR AND THERMOMETER IN WARM WATER. SWITCH ON TRANSMITTER AND RECEIVER. COUNT NUMBER OF CLICKS IN 15 SECONDS AND RECORD COUNT AND TEMPERATURE. REPEAT AS WATER COOLS. ADD ICE FOR COLD TEMPERATURES. SAMPLE CALIBRATION:

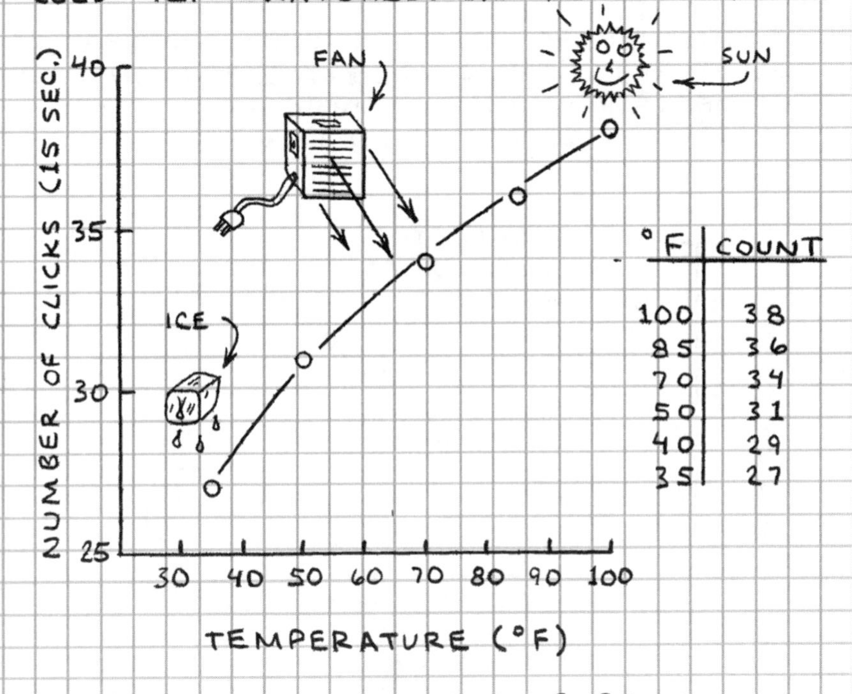

°F	COUNT
100	38
85	36
70	34
50	31
40	29
35	27

CALIBRATION WILL CHANGE IF R1 IS READJUSTED. OK TO USE FIXED RESISTOR FOR R1. CIRCUIT WORKS BEST WITH THERMISTOR THAT HAS A RESISTANCE AT ROOM TEMPERATURE (25°C) OF 10K.

LED TELEMETRY TRANSMITTER

THIS LED FLASHER WILL TELL YOU THE TEMPERATURE AT ITS LOCATION FROM ANYWHERE YOU CAN SEE ITS FLASHES. CHECK TEMPERATURE OF GREENHOUSE, GARDEN, ETC. WHILE YOU STAY INDOORS. WORKS BEST IN SUBDUED LIGHT.

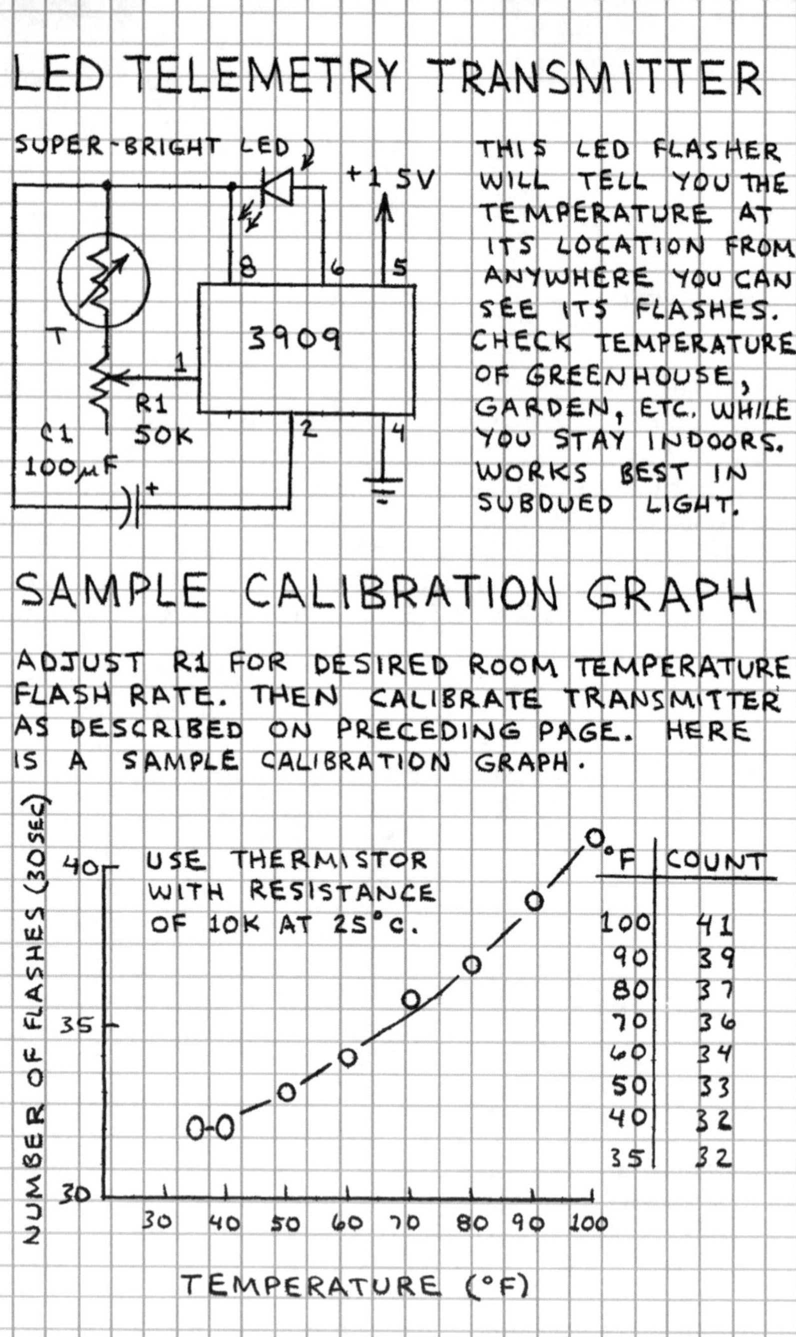

SAMPLE CALIBRATION GRAPH

ADJUST R1 FOR DESIRED ROOM TEMPERATURE FLASH RATE. THEN CALIBRATE TRANSMITTER AS DESCRIBED ON PRECEDING PAGE. HERE IS A SAMPLE CALIBRATION GRAPH.

°F	COUNT
100	41
90	39
80	37
70	36
60	34
50	33
40	32
35	32

CALIBRATION IS FOR NUMBER OF FLASHES IN 30 SECONDS. YOU CAN MAKE A MORE ACCURATE GRAPH BY COUNTING FLASHES IN 60 SECONDS. R1 CAN BE FIXED RESISTOR.

ELECTRONIC CRICKET

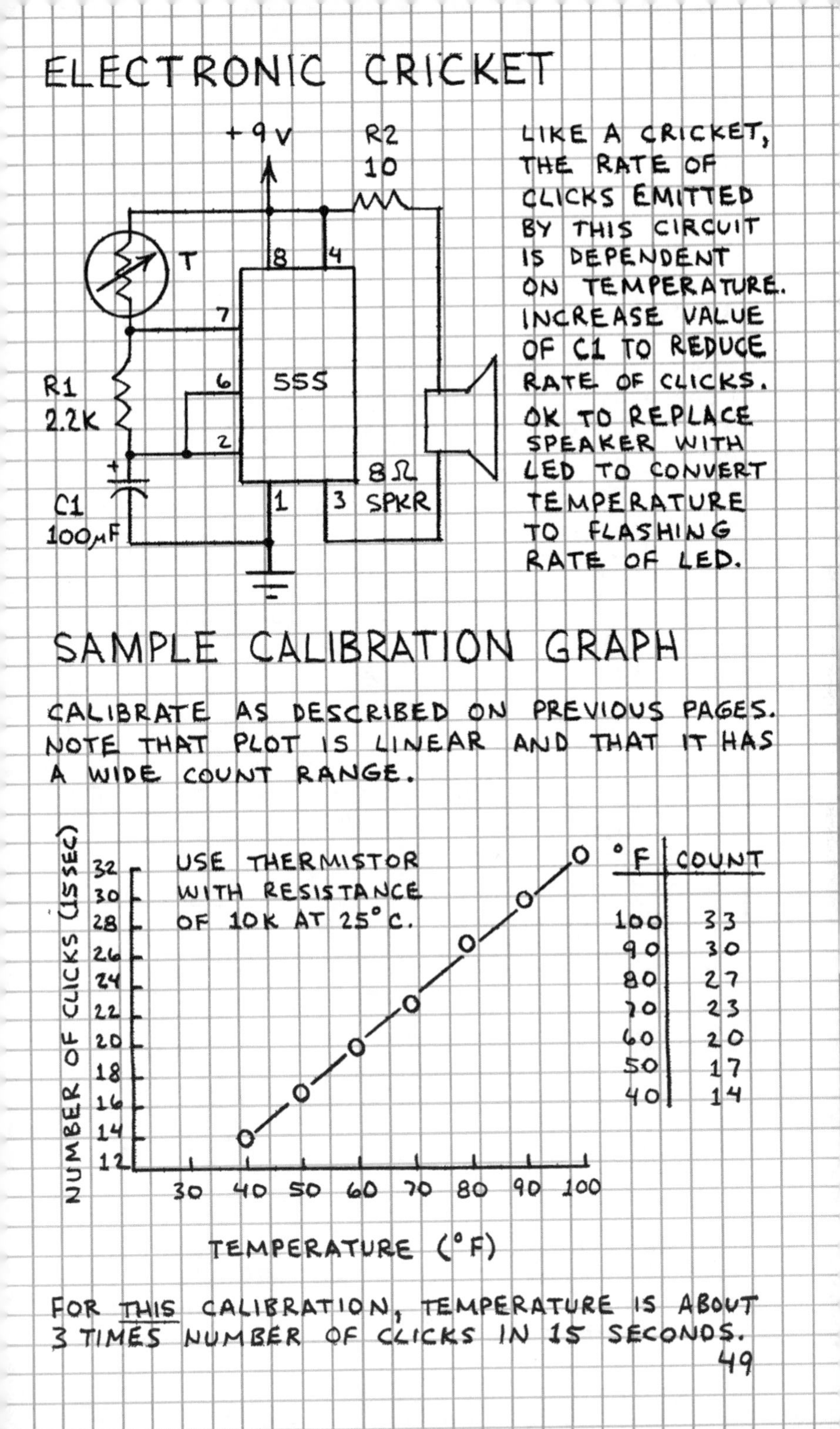

LIKE A CRICKET, THE RATE OF CLICKS EMITTED BY THIS CIRCUIT IS DEPENDENT ON TEMPERATURE. INCREASE VALUE OF C1 TO REDUCE RATE OF CLICKS. OK TO REPLACE SPEAKER WITH LED TO CONVERT TEMPERATURE TO FLASHING RATE OF LED.

SAMPLE CALIBRATION GRAPH

CALIBRATE AS DESCRIBED ON PREVIOUS PAGES. NOTE THAT PLOT IS LINEAR AND THAT IT HAS A WIDE COUNT RANGE.

°F	COUNT
100	33
90	30
80	27
70	23
60	20
50	17
40	14

FOR THIS CALIBRATION, TEMPERATURE IS ABOUT 3 TIMES NUMBER OF CLICKS IN 15 SECONDS.

ANALOG DATA LOGGER

YOU CAN RECORD EXPERIMENTAL DATA ON MAGNETIC TAPE WITH THE HELP OF A SMALL CASSETTE TAPE RECORDER. FIRST, CONVERT SIGNAL TO BE SAVED INTO A VOLTAGE. THEN TRANSFORM THE VOLTAGE INTO AN AUDIO-FREQUENCY TONE WITH A VOLTAGE-TO-FREQUENCY (V/F) CONVERTER. RECORD TONE ON MAGNETIC TAPE. RETRIEVE DATA BY PLAYING TAPE THROUGH A FREQUENCY-TO-VOLTAGE (F/V) CONVERTER.

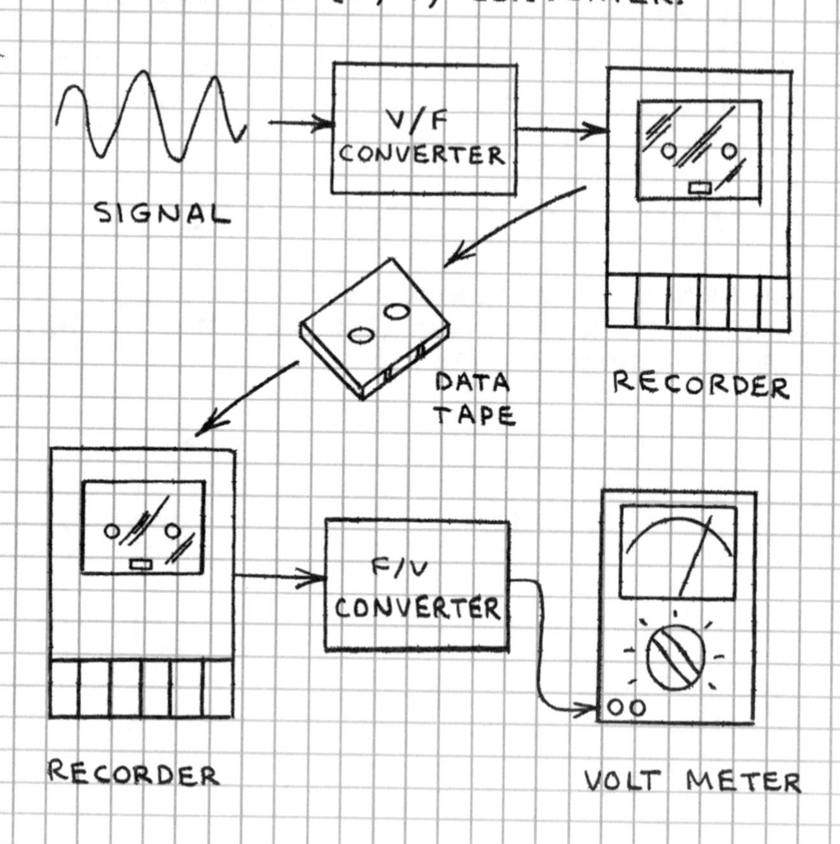

FOR BEST RESULTS, USE QUALITY RECORDING TAPE. BETTER QUALITY RECORDERS WORK BEST. YOU CAN SQUEEZE MORE DATA ON A TAPE BY RECORDING 5 SECOND "SNAP SHOTS."

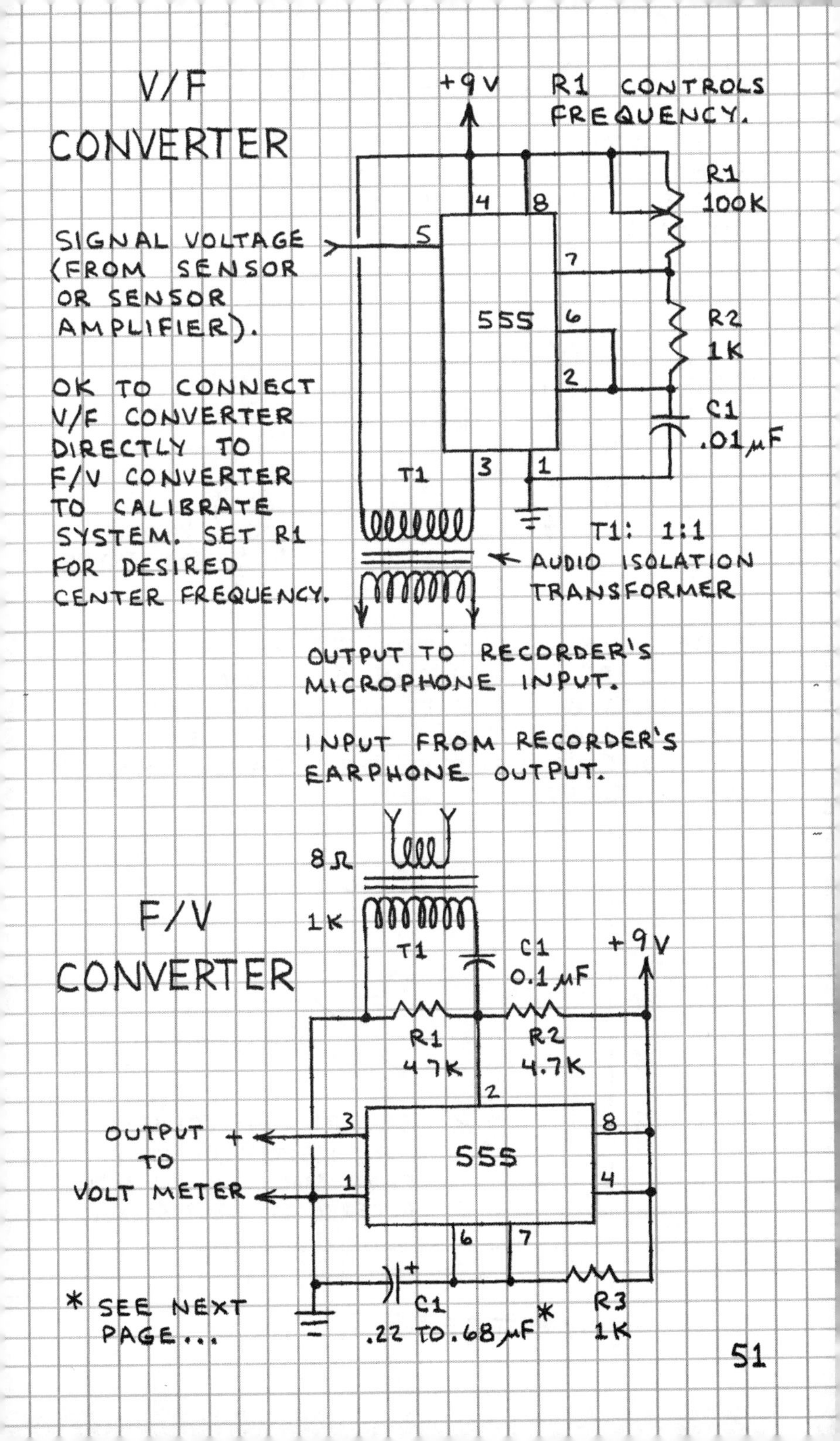
V/F
CONVERTER
+9V
R1 CONTROLS
FREQUENCY.
R1
100K
4
8
5
SIGNAL VOLTAGE
(FROM SENSOR
OR SENSOR
AMPLIFIER).
7
555
6
R2
1K
2
C1
.01μF
OK TO CONNECT
V/F CONVERTER
DIRECTLY TO
F/V CONVERTER
TO CALIBRATE
SYSTEM. SET R1
FOR DESIRED
CENTER FREQUENCY.
T1
3
1
T1: 1:1
AUDIO ISOLATION
TRANSFORMER
OUTPUT TO RECORDER'S
MICROPHONE INPUT.
INPUT FROM RECORDER'S
EARPHONE OUTPUT.
8Ω
1K
F/V
CONVERTER
T1
C1
0.1μF
+9V
R1
47K
R2
4.7K
2
3
8
OUTPUT +
TO
VOLT METER
555
1
4
6
7
* SEE NEXT
PAGE...
C1
.22 TO .68μF*
R3
1K

DATA LOGGER OPERATION

THE OUTPUT FROM MOST SENSORS CAN BE CHANGED TO A VOLTAGE. FOR EXAMPLE, THESE CIRCUITS BOTH CHANGE LIGHT INTENSITY INTO A VARIABLE VOLTAGE:

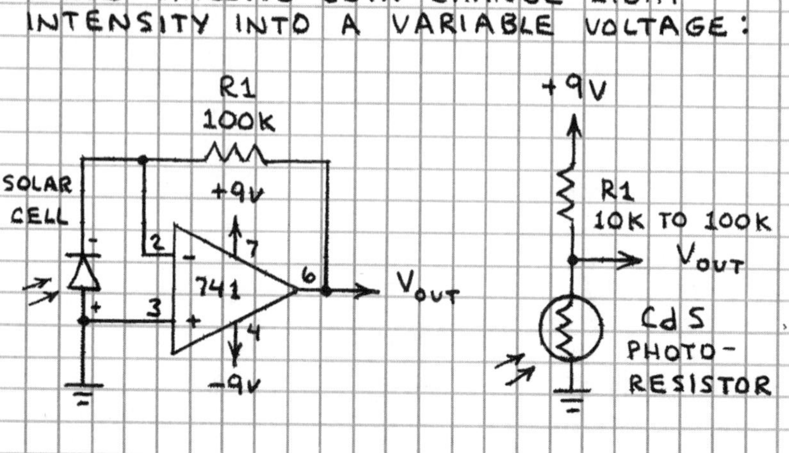

HERE IS A CALIBRATION GRAPH FOR TWO VALUES OF C1 IN THE F/V CONVERTER. THE GRAPH SHOULD BE CONSIDERED AS APPROXIMATE SINCE DIFFERENCES IN COMPONENT VALUES WILL CAUSE CHANGES IN THE GRAPH.

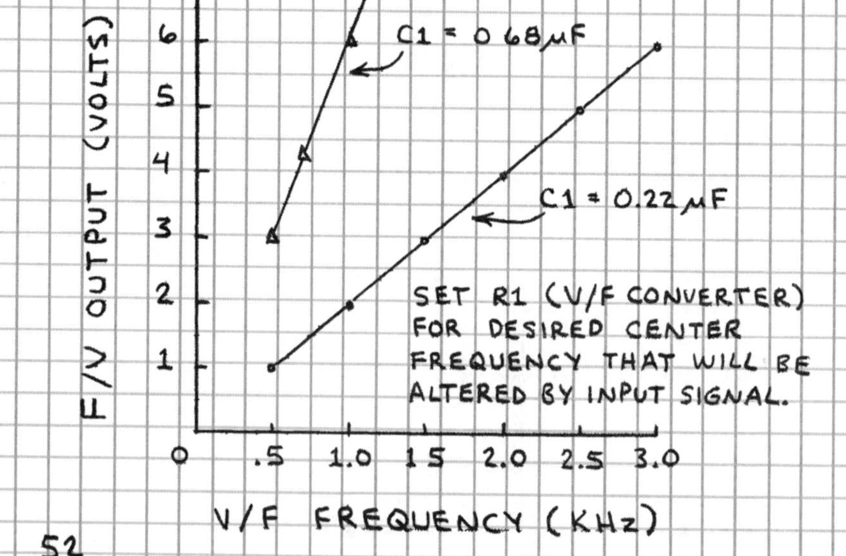

II. ENVIRONMENTAL SCIENCE

OVERVIEW

NATURAL ENVIRONMENTS CONSTANTLY CHANGE IN RESPONSE TO MANY INFLUENCES. FOR EXAMPLE:

- ☐ SUBTLE CHANGES IN THE SUN'S ENERGY MAY CAUSE MAJOR CLIMATE CHANGES ON EARTH.
- ☐ MAJOR VOLCANOES CAN EJECT SULFUR DIOXIDE (SO_2) INTO THE ATMOSPHERE. THE SO_2 COMBINES WITH WATER VAPOR TO FORM A MIST OF SULFURIC ACID (H_2SO_4) WHICH BLOCKS SUNLIGHT.
- ☐ INSECTS CAN DESTROY LARGE STANDS OF PLANTS AND EVEN TREES.
- ☐ A BEAVER DAM CAN CREATE A LARGE POND THAT ALTERS THE POPULATION OF PLANTS AND ANIMALS.
- ☐ EMISSIONS FROM COAL-FIRED POWER PLANTS CAN COMBINE WITH WATER VAPOR TO FORM THICK BLANKETS OF HAZE.

THE PROJECTS THAT FOLLOW DESCRIBE THE BASICS OF WATER TESTING AND MEASURING SOUND, HAZE, TEMPERATURE, SUNLIGHT AND LIGHTNING. BY REGULARLY MONITORING ONE OR MORE OF THESE OR OTHER PARAMETERS, YOU CAN MAKE AN IMPORTANT CONTRIBUTION TO ENVIRONMENTAL SCIENCE.

SAFETY

ALWAYS USE CAUTION WHEN MEASURING THE ENVIRONMENT, ESPECIALLY DURING LIGHTNING STORMS AND AROUND BODIES OF WATER. USE EAR PROTECTORS WHEN MEASURING LOUD SOUND. NEVER LOOK AT THE SUN WHEN MEASURING ITS LIGHT.

GRAPHING YOUR DATA

ONE OF THE BEST WAYS TO PRESENT YOUR DATA IS TO PLOT IT ON A GRAPH. THESE GRAPHS SHOW MY OBSERVATIONS AT GERONIMO CREEK, TEXAS.

LINE GRAPH

ALLOWS YOU TO SEE CHANGES IN TRENDS.

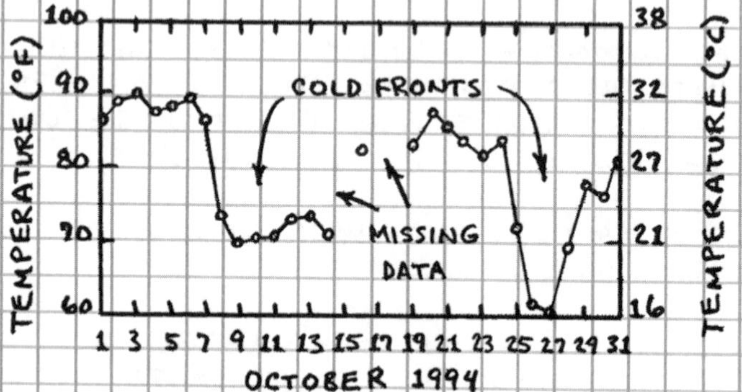

HISTOGRAM

A BARGRAPH THAT SHOWS THE FREQUENCY OF OCCURRENCE IS A HISTOGRAM. THIS ONE SHOWS THE CLASSIC BELL-SHAPED CURVE.

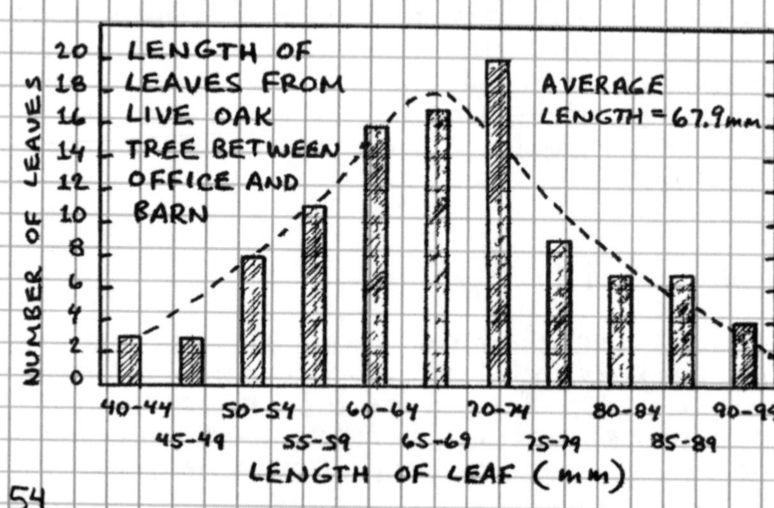

SCATTER GRAPH

IS THERE A RELATIONSHIP BETWEEN TWO SETS OF DATA? ASSIGN ONE SET TO X AXIS (↔) AND THE OTHER TO Y AXIS (↕). PLOT PAIRS OF DATA AS POINTS. THE MORE CLOSELY THE POINTS ARE CLUSTERED ALONG A LINE, THE BETTER THE CORRELATION OR AGREEMENT OF THE TWO SETS OF DATA.

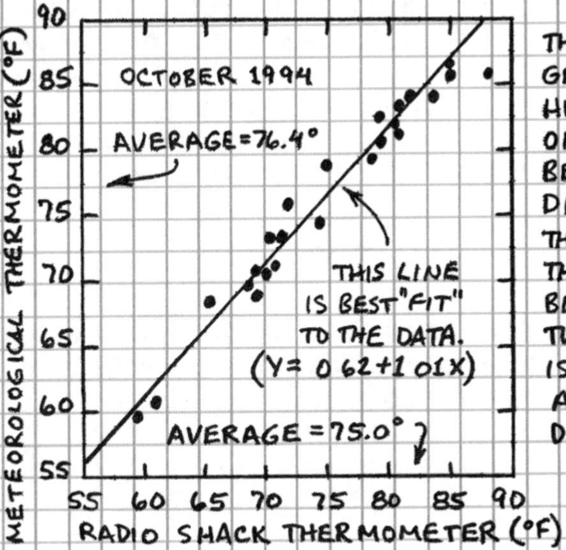

THIS SCATTER GRAPH SHOWS A HIGH DEGREE OF CORRELATION BETWEEN TWO DIGITAL THERMOMETERS. THE DIFFERENCE BETWEEN THE TWO AVERAGES IS THE "OFFSET," A CONSISTENT DIFFERENCE.

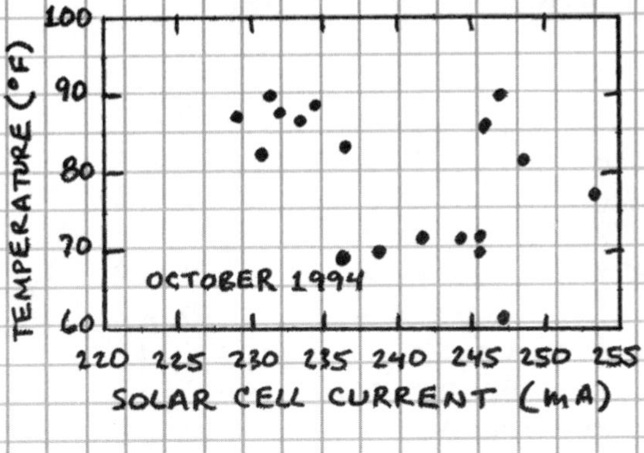

THIS SCATTER GRAPH SHOWS NO OBVIOUS CORRELATION BETWEEN TEMPERATURE AND INTENSITY OF SUN LIGHT.

GOING FURTHER

FOR SERIOUS ANALYSIS, USE A SCIENTIFIC CALCULATOR OR COMPUTER SPREADSHEET TO GRAPH YOUR DATA.

SOUND

WHEN YOU HEAR A SOUND, YOUR EARS ARE RESPONDING TO TINY, RAPID CHANGES IN THE PRESSURE OF THE AIR. THESE CHANGES ARE SOUND WAVES. THEY MAY HAVE A SINGLE PITCH (FREQUENCY) AND CONSTANT LOUDNESS (INTENSITY OR AMPLITUDE). OR THEY MAY BE A COMPLEX MIXTURE OF WAVES WITH DIFFERENT FREQUENCIES AND AMPLITUDES. REPETITIOUS WAVES OF UNIFORM OR GRADUALLY CHANGING FREQUENCY AND AMPLITUDE ARE USUALLY MORE PLEASANT THAN IRREGULAR, ABRUPTLY CHANGING WAVES.

SOUND INTENSITY

SINCE THE EAR RESPONDS TO AN ENORMOUS RANGE OF SOUND LEVELS, THE INTENSITY OF SOUND IS EXPRESSED USING A LOGARITHMIC SCALE IN WHICH 0 DECIBELS IS A BARELY PERCEPTIBLE SOUND WITH AN INTENSITY OF 10^{-12} WATTS PER SQUARE METER (W/m^2).

RATIO OF MEASURED TO REFERENCE SOUND	RATIO IN DECIBELS
1	0 dB
10	10 dB
100	20 dB
1,000	30 dB
10,000	40 dB
100,000	50 dB
1,000,000	60 dB
10,000,000	70 dB
100,000,000	80 dB
1,000,000,000	90 dB
10,000,000,000	100 dB
100,000,000,000	110 dB
1,000,000,000,000	120 dB
10,000,000,000,000	130 dB
100,000,000,000,000	140 dB

A 10dB INCREASE IS AN INCREASE OF 10 TIMES THE INITIAL INTENSITY. (60dB = 10 × 50dB.)

SOUND FREQUENCY

SOUND WAVES RANGE FROM PURE SINE WAVES TO COMPLEX COMBINATIONS OF WAVES. THIS SINE WAVE HAS A FREQUENCY OF 1 CYCLE PER SECOND (1 HERTZ OR 1 Hz):

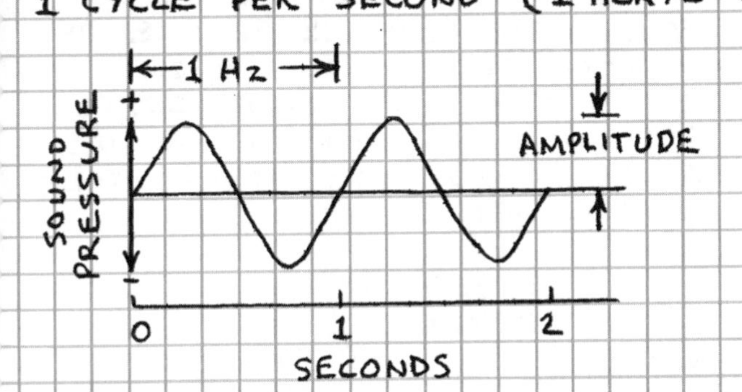

RANGE OF HUMAN HEARING

THE NORMAL HUMAN EAR CAN PERCEIVE SOUNDS RANGING IN FREQUENCY FROM 20 TO 20,000 Hz. THE PERCEPTION OF HIGH FREQUENCIES DECREASES WITH AGE AND IS REDUCED BY REPEATED EXPOSURE TO VERY LOUD SOUNDS. INFRASONIC SOUND IS SOUND HAVING A FREQUENCY BELOW THE RANGE OF HUMAN HEARING. ULTRASONIC SOUND IS SOUND HAVING A FREQUENCY ABOVE THE RANGE OF HUMAN HEARING.

THE SPEED OF SOUND

THE SPEED OF SOUND IN DRY AIR AT 0° CELSIUS (32° FAHRENHEIT) IS 331 METERS (1086 FEET) PER SECOND. THE SPEED INCREASES WITH TEMPERATURE. AT 20° C (68° F), THE SPEED OF SOUND IN AIR IS 343 METERS (1125 FEET) PER SECOND. SOUND WAVES TRAVEL THROUGH LIQUIDS AND SOLIDS MUCH MORE RAPIDLY THAN THROUGH AIR. THE SPEED OF SOUND IN WATER AT 25°C (77°F) IS 1497 METERS (4911 FEET) PER SECOND.

MEASURING SOUND INTENSITY

RADIO SHACK SOUND LEVEL METERS ARE IDEAL FOR CONDUCTING SOUND SURVEYS. WHEN MEASURING SOUND COMING FROM ONE DIRECTION, DO NOT HOLD THE METER BETWEEN YOUR BODY AND THE SOURCE OF THE SOUND. HOLD THE METER TO ONE SIDE AND POINT IT AT THE SOUND SOURCE. USE FAST RESPONSE FOR SPORADIC SOUNDS OR TO MEASURE PEAKS. USE SLOW RESPONSE TO MEASURE AVERAGE SOUND LEVEL.

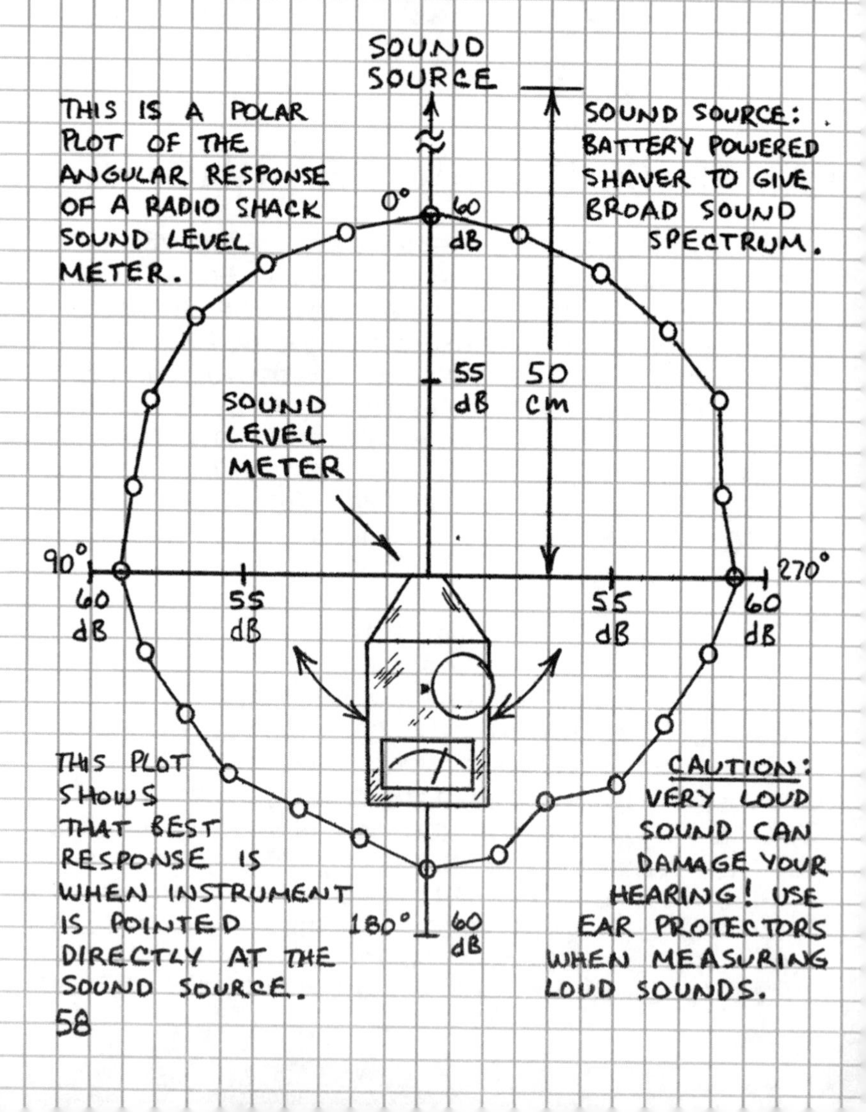

TYPICAL SOUND LEVELS

SOUND INTENSITY CAN VARY WITH WIND AND LOCATION OF THE SOUND LEVEL METER. HERE ARE SOME TYPICAL LEVELS:

SOURCE	INTENSITY (dB)
JET AIRCRAFT (6m)	140
THRESHOLD OF PAIN	130
SUBWAY TRAIN	102
NIAGARA FALLS	92
PASSING TRUCK (6m)	80
PIANO (EAR OF PLAYER)	80
WATER FILLING TUB (1m)	76
VACUUM CLEANER (2m)	72
TYPICAL CAR (5m)	70
JET AIRCRAFT (2KM)	68
EXHAUST FAN (2m)	68
COMPUTER (1m)	58
RADIO (3m)	57
TYPICAL OFFICE	55
TYPICAL RESIDENCE	40
WHISPER (1.5m)	18
THRESHOLD OF SOUND	0

ARTIFICIAL SOUND SOURCES

ARTIFICIAL SOUND SOURCES CAN BE USEFUL IN EVALUATING THE ACOUSTICAL PROPERTIES OF A ROOM OR AUDITORIUM. THEY ARE ESPECIALLY USEFUL WHEN USED WITH A SOUND LEVEL METER. SMALL ELECTRIC MOTORS AND ELECTRIC SHAVERS CAN BE USED AS BROAD BAND, LOW FREQUENCY SOUND SOURCES. THE CIRCUITS BELOW ARE TONE SOURCES.

SINGLE FREQUENCY TONE

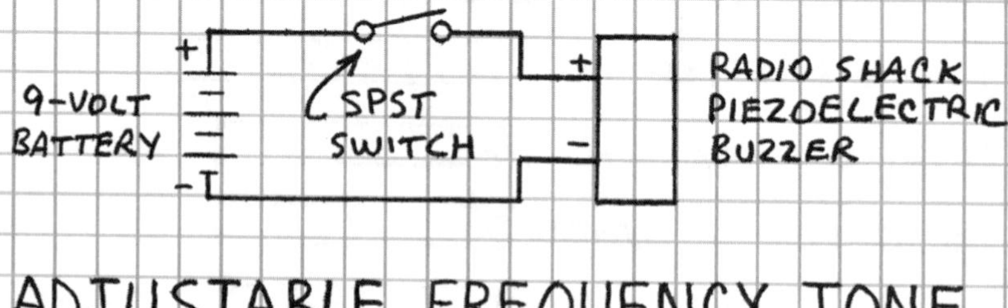

ADJUSTABLE FREQUENCY TONE

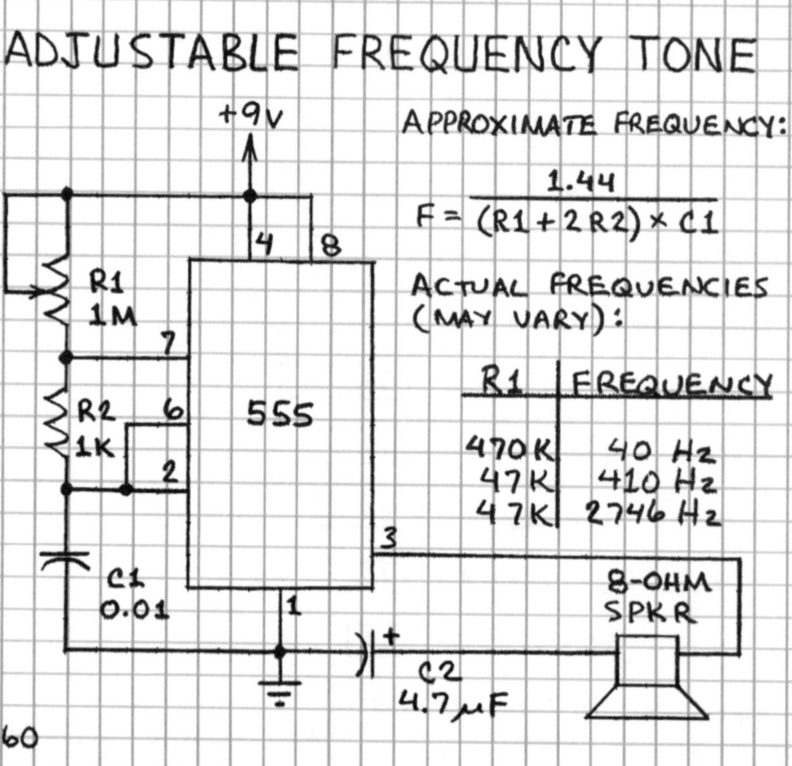

SOUND INTENSITY STUDIES

THE INTENSITY OF A SOUND IS INVERSELY PROPORTIONAL TO THE SQUARE OF THE DISTANCE TO THE SOURCE. THEREFORE A PLOT OF SOUND INTENSITY IN DECIBELS VS. DISTANCE TO THE SOURCE FORMS A STRAIGHT LINE.

RADIO SHACK PIEZOELECTRIC BUZZER

STRAIGHT LINE

SOUND LEVEL (dB)

120
110
100
90
80
70

0 1 2 3 4 5

DISTANCE (METERS)

BATTERY-POWERED ELECTRIC SHAVER

STRAIGHT LINE

BACKGROUND SOUND

SOUND LEVEL (dB)

78
74
70
66
62
58
54

10 20 30 40 50 60 70 80 90 100 110 120

DISTANCE (CENTIMETERS)

8-CYLINDER CAR ENGINE

STRAIGHT LINE

BACKGROUND SOUND

SOUND LEVEL (dB)

90
80
70
60

0 1 2 3 4 5 6 7

DISTANCE (METERS)

NOTE THAT STRAIGHT LINE BEGINS AWAY FROM SOURCE. TO ESTIMATE THE SOUND INTENSITY 10 METERS FROM A LOUD BUT DISTANT WATERFALL, JET, TRAIN, BAND, ETC., MAKE SEVERAL MEASUREMENTS AT DIFFERENT DISTANCES. PLOT THE DATA AND DRAW A LINE THROUGH THE POINTS. EXTEND THE LINE TO ESTIMATE THE INTENSITY OF THE SOUND NEARER ITS SOURCE.

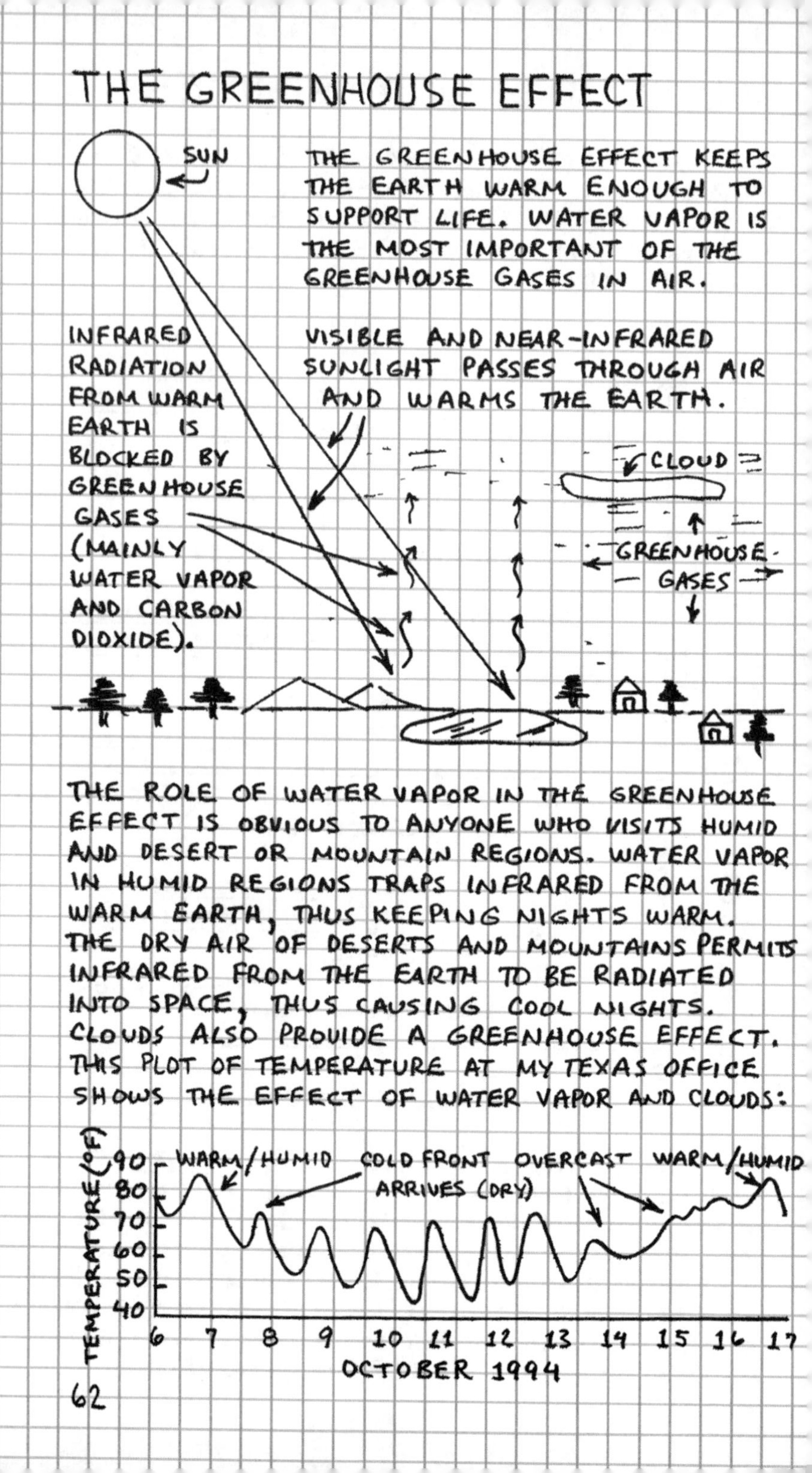
THE GREENHOUSE EFFECT
SUN
THE GREENHOUSE EFFECT KEEPS THE EARTH WARM ENOUGH TO SUPPORT LIFE. WATER VAPOR IS THE MOST IMPORTANT OF THE GREENHOUSE GASES IN AIR.
INFRARED RADIATION FROM WARM EARTH IS BLOCKED BY GREENHOUSE GASES (MAINLY WATER VAPOR AND CARBON DIOXIDE).
VISIBLE AND NEAR-INFRARED SUNLIGHT PASSES THROUGH AIR AND WARMS THE EARTH.
CLOUD
GREENHOUSE GASES
THE ROLE OF WATER VAPOR IN THE GREENHOUSE EFFECT IS OBVIOUS TO ANYONE WHO VISITS HUMID AND DESERT OR MOUNTAIN REGIONS. WATER VAPOR IN HUMID REGIONS TRAPS INFRARED FROM THE WARM EARTH, THUS KEEPING NIGHTS WARM. THE DRY AIR OF DESERTS AND MOUNTAINS PERMITS INFRARED FROM THE EARTH TO BE RADIATED INTO SPACE, THUS CAUSING COOL NIGHTS. CLOUDS ALSO PROVIDE A GREENHOUSE EFFECT. THIS PLOT OF TEMPERATURE AT MY TEXAS OFFICE SHOWS THE EFFECT OF WATER VAPOR AND CLOUDS:
TEMPERATURE (°F)
90
80
70
60
50
40
WARM/HUMID
COLD FRONT ARRIVES (DRY)
OVERCAST
WARM/HUMID
6
7
8
9
10
11
12
13
14
15
16
17
OCTOBER 1994

WATER VAPOR

THE ATMOSPHERE ALWAYS INCLUDES SOME WATER VAPOR. AIR IS NOT A CONTAINER FOR WATER; WATER MOLECULES ARE PART OF THE AIR. WATER VAPOR CAN BE UP TO 4% OF WARM, TROPICAL AIR. COLD AIR IS MUCH DRIER, AND AT -40°C (-40°F) THE MAXIMUM PERCENTAGE OF WATER IN AIR CANNOT BE GREATER THAN ABOUT 0.02%.

RELATIVE HUMIDITY

RELATIVE HUMIDITY IS THE RATIO OF THE ACTUAL TO THE MAXIMUM POSSIBLE WATER VAPOR IN THE AIR AT A GIVEN TEMPERATURE. SINCE THE MAXIMUM POSSIBLE WATER VAPOR IN WARM AIR IS MUCH HIGHER THAN THAT IN COLD AIR, RELATIVE HUMIDITY IS DEPENDENT ON TEMPERATURE. THUS THE RELATIVE HUMIDITY ON A COOL SPRING MORNING CAN BE 95% AND ONLY 50% LATER IN THE DAY, EVEN THOUGH THE TOTAL WATER VAPOR IN THE AIR HAS NOT CHANGED.

MEASURING RELATIVE HUMIDITY

USE A RELATIVE HUMIDITY METER. OR USE TWO THERMOMETERS, ONE WITH A <u>WET</u> SENSOR OR BULB. BLOW AIR PAST THE <u>WET</u> SENSOR FOR A MINUTE. THEN USE CHART ON FOLLOWING TWO PAGES TO FIND RELATIVE HUMIDITY.

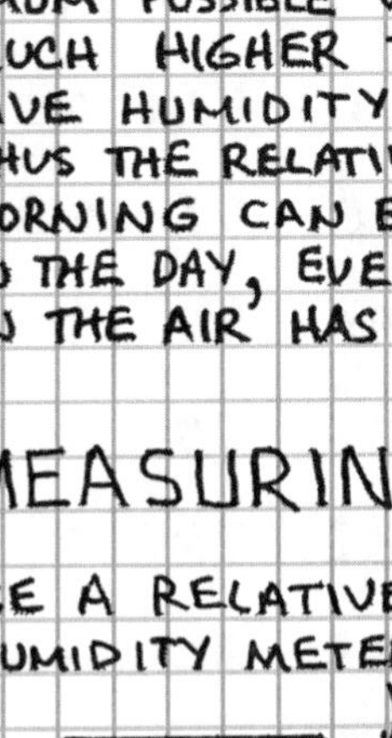

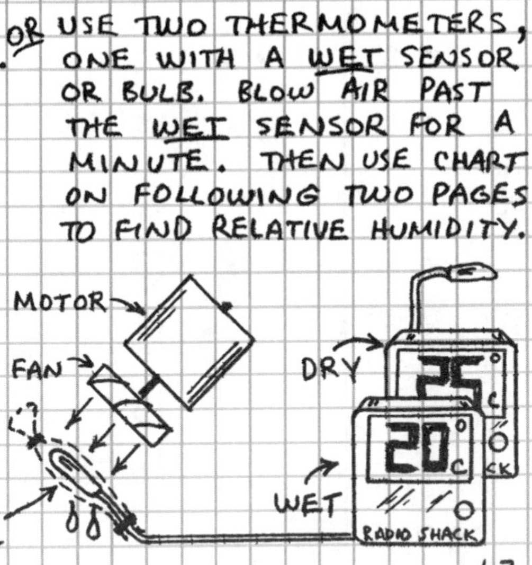

RELATIVE HUMIDITY (%)

DRY BULB TEMPERATURE (°C)	DRY BULB (°C) − WET BULB (°C)					
	0.5	1.0	1.5	2.0	2.5	3.0
-5.0	88	77	66	54	43	32
-2.5	90	80	70	60	50	41
0.0	91	82	73	65	56	47
2.5	92	84	76	68	61	53
5.0	93	86	78	71	65	58
7.5	93	87	80	74	68	62
10.0	94	88	82	76	71	65
12.5	94	89	84	78	73	68
15.0	95	90	85	80	75	70
17.5	95	90	86	81	77	72
20.0	95	91	87	82	78	74
22.5	96	92	87	83	80	76
25.0	96	92	88	84	81	77
27.5	96	92	89	85	82	78
30.0	96	93	89	86	82	79
32.5	97	93	90	86	83	80
35.0	97	93	90	87	84	81
37.5	97	94	91	87	85	82
40.0	97	94	91	88	85	82

DRY BULB IS TEMPERATURE OF THE AIR.
WET BULB IS TEMPERATURE OF VENTILATED SENSOR WRAPPED IN MOIST CLOTH.

3.5	4.0	4.5	5.0	7.5	10.0	12.5	15.0	17.5
21	11	0						
31	22	12	3					
39	31	23	15					
46	38	31	24					
51	45	38	32	1				
56	50	44	38	11				
60	54	49	44	19				
63	58	53	48	25	4			
66	61	57	52	31	12			
68	64	60	55	36	18	2		
70	66	62	58	40	24	8		
72	68	64	61	44	28	14	1	
73	70	66	63	47	32	19	7	
75	71	68	65	50	36	23	12	1
76	73	70	67	52	39	27	16	6
77	74	71	68	54	42	30	20	11
78	75	72	69	56	44	33	23	14
79	76	73	70	58	46	36	26	18
79	77	74	72	59	48	38	29	21

SOURCE:
"METEOROLOGY" BY J. MORAN AND M. MORGAN, MACMILLAN PUBLISHING CO., p. 560 (1989).

TO CONVERT °C TO °FAHRENHEIT:

°F = (°C × 9/5) + 32

EXAMPLE:
DRY = 25°C
WET = 20°C
DRY − WET = 5°C
RH = 63%

THE HEAT ISLAND EFFECT

TOWNS AND CITIES ARE SOMETIMES CALLED "HEAT ISLANDS" SINCE THEY ARE GENERALLY WARMER THAN THE NEARBY COUNTRYSIDE. YOU CAN EASILY MEASURE YOUR CITY'S HEAT ISLAND EFFECT WHILE DRIVING ACROSS TOWN. YOU WILL NEED:

- ☐ A NOTEBOOK OR TAPE RECORDER TO RECORD YOUR MEASUREMENTS.
- ☐ A THERMOMETER. (DIGITAL TYPE WITH SENSOR ON A CABLE WORKS BEST.)
- ☐ A FRIEND OR RELATIVE TO DRIVE WHILE YOU RECORD DATA. CAUTION: DO NOT ATTEMPT TO DRIVE AND RECORD DATA!

RECORD THE FOLLOWING:
1. DATE
2. START TIME
3. LOCATION
4. ODOMETER
5. TEMPERATURE
6. END TIME

START LIFEGATE SCHOOL 0708:18

GUADALUPE RIVER BRIDGE

RADIO SHACK STORE

CEDAR STREET

JUNCTION 466/477

FOG PATCH

COURT STREET

THE TEMPERATURE AT START AND STOP POINTS WAS EQUAL AT 0706.

TEMPERATURE (FAHRENHEIT): 73° 72° 71° 70° 69° 68° 67° 66° 65° 64°

0 1(1.6) 2(3.2) 3(4.8) 4(6.4)

ODOMETER MILES

THE TEMPERATURE SENSOR MUST BE SHIELDED FROM SUNLIGHT AND KEPT AWAY FROM THE CAR'S ENGINE AND EXHAUST. MAKE HOLLOW TUBE FROM STIFF WHITE PAPER, TAPE TO SIDE MIRROR OR DOOR HANDLE WITH OPEN END FACING FORWARD. TAPE SENSOR INSIDE TUBE.

GOING FURTHER: MEASURE HEAT ISLAND EFFECT AT DIFFERENT TIMES OF DAY AND YEAR. IS THE EFFECT GREATER IN SUMMER OR WINTER? MORNING OR NIGHT? CAN YOU MEASURE THE HEAT ISLAND EFFECT OF LARGE PARKING LOTS HEATED BY THE SUN, FACTORIES, SUBDIVISIONS, ETC.?

GRAPH YOUR DATA — LIKE THIS

ACROSS TOWN FROM SARAH'S SCHOOL TO HOME
OCTOBER 4, 1994
FORREST MIMS

INTERSTATE 10
MOTEL
JUNCTION 123/123 BUSINESS
RAILROAD OVERPASS
MARTINDALE ROAD
ONE MILE ROAD
DIP
STOP HOME 0726'10

HEAT FROM RISING SUN CAUSED INCREASE IN TEMPERATURE AT HOME AT 0726.

23° 22° 21° 20° 19° 18°
TEMPERATURE (CELSIUS)

5 (8) 6 (9.7) 7 (11.3) 8 (12.9) 9 (14.5) 10 (16.1)
(KILOMETERS IN PARENTHESES)

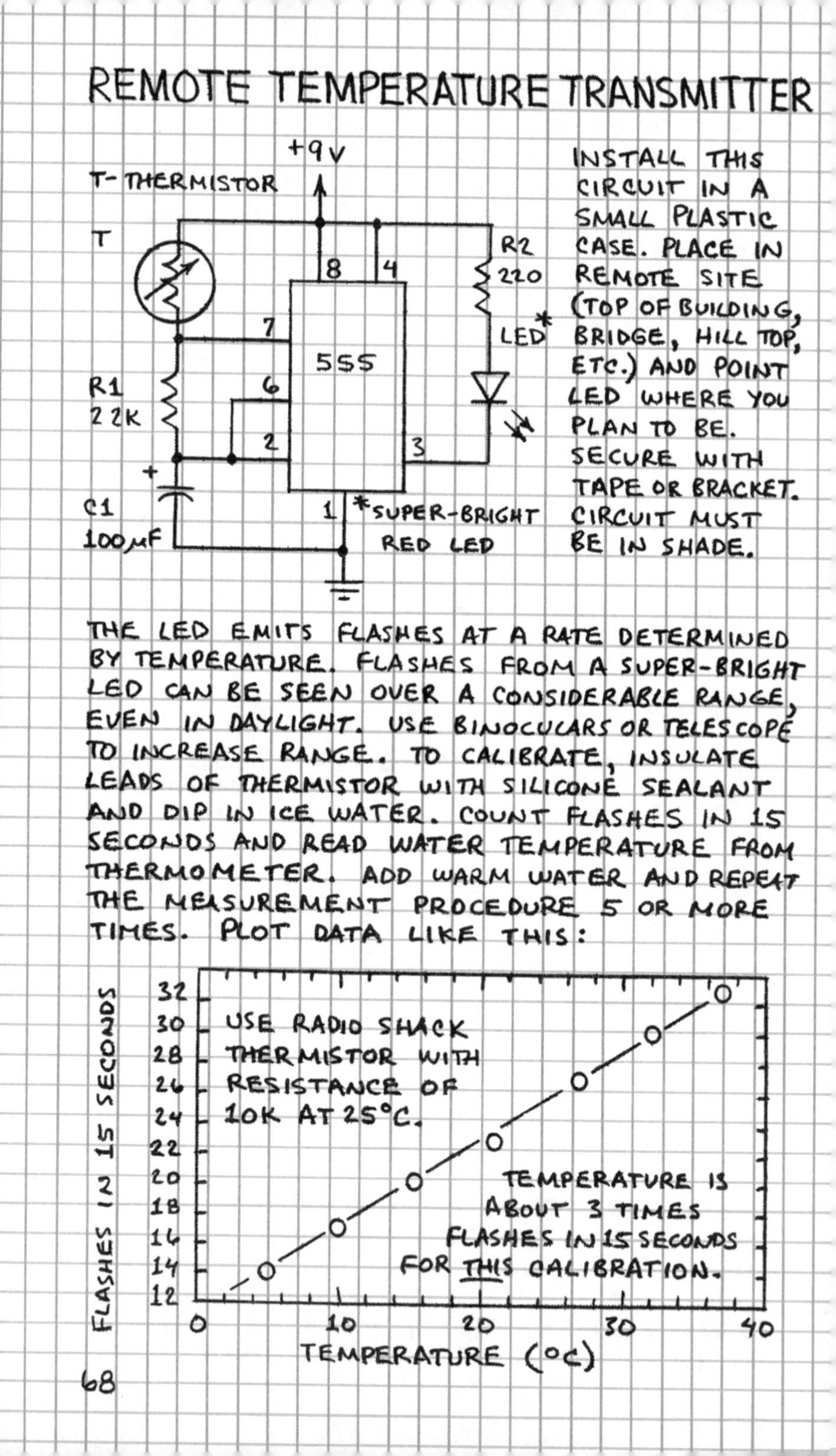

REMOTE TEMPERATURE TRANSMITTER

INSTALL THIS CIRCUIT IN A SMALL PLASTIC CASE. PLACE IN REMOTE SITE (TOP OF BUILDING, BRIDGE, HILL TOP, ETC.) AND POINT LED WHERE YOU PLAN TO BE. SECURE WITH TAPE OR BRACKET. CIRCUIT MUST BE IN SHADE.

THE LED EMITS FLASHES AT A RATE DETERMINED BY TEMPERATURE. FLASHES FROM A SUPER-BRIGHT LED CAN BE SEEN OVER A CONSIDERABLE RANGE, EVEN IN DAYLIGHT. USE BINOCULARS OR TELESCOPE TO INCREASE RANGE. TO CALIBRATE, INSULATE LEADS OF THERMISTOR WITH SILICONE SEALANT AND DIP IN ICE WATER. COUNT FLASHES IN 15 SECONDS AND READ WATER TEMPERATURE FROM THERMOMETER. ADD WARM WATER AND REPEAT THE MEASUREMENT PROCEDURE 5 OR MORE TIMES. PLOT DATA LIKE THIS:

ABOVE-BELOW TEMPERATURE RECORDER

SOME FRUIT TREES REQUIRE A MINIMUM NUMBER OF HOURS WHEN THE TEMPERATURE IS BELOW FREEZING. THIS CIRCUIT RECORDS THE TIME THE TEMPERATURE IS BELOW 0°C (32°F) OR ANOTHER TEMPERATURE SELECTED BY R2.

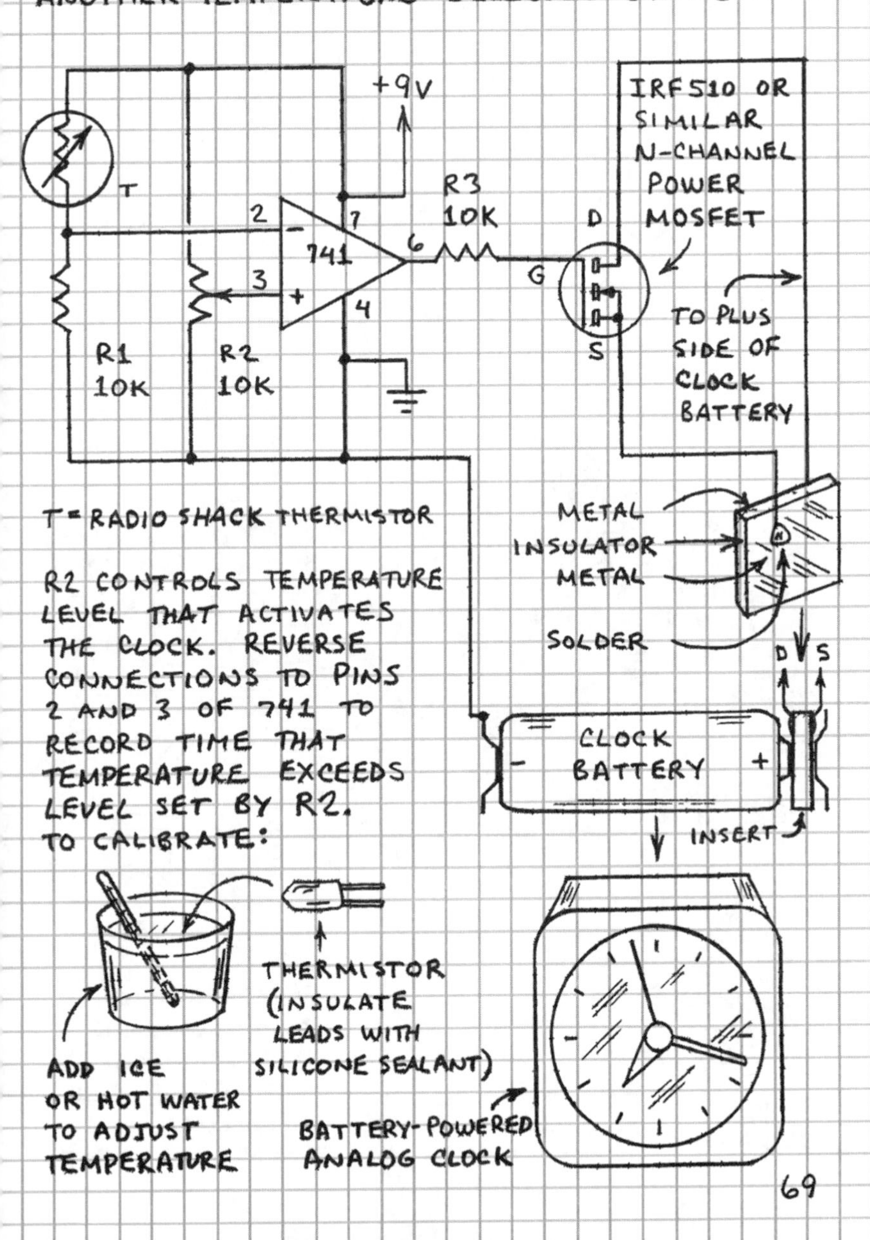

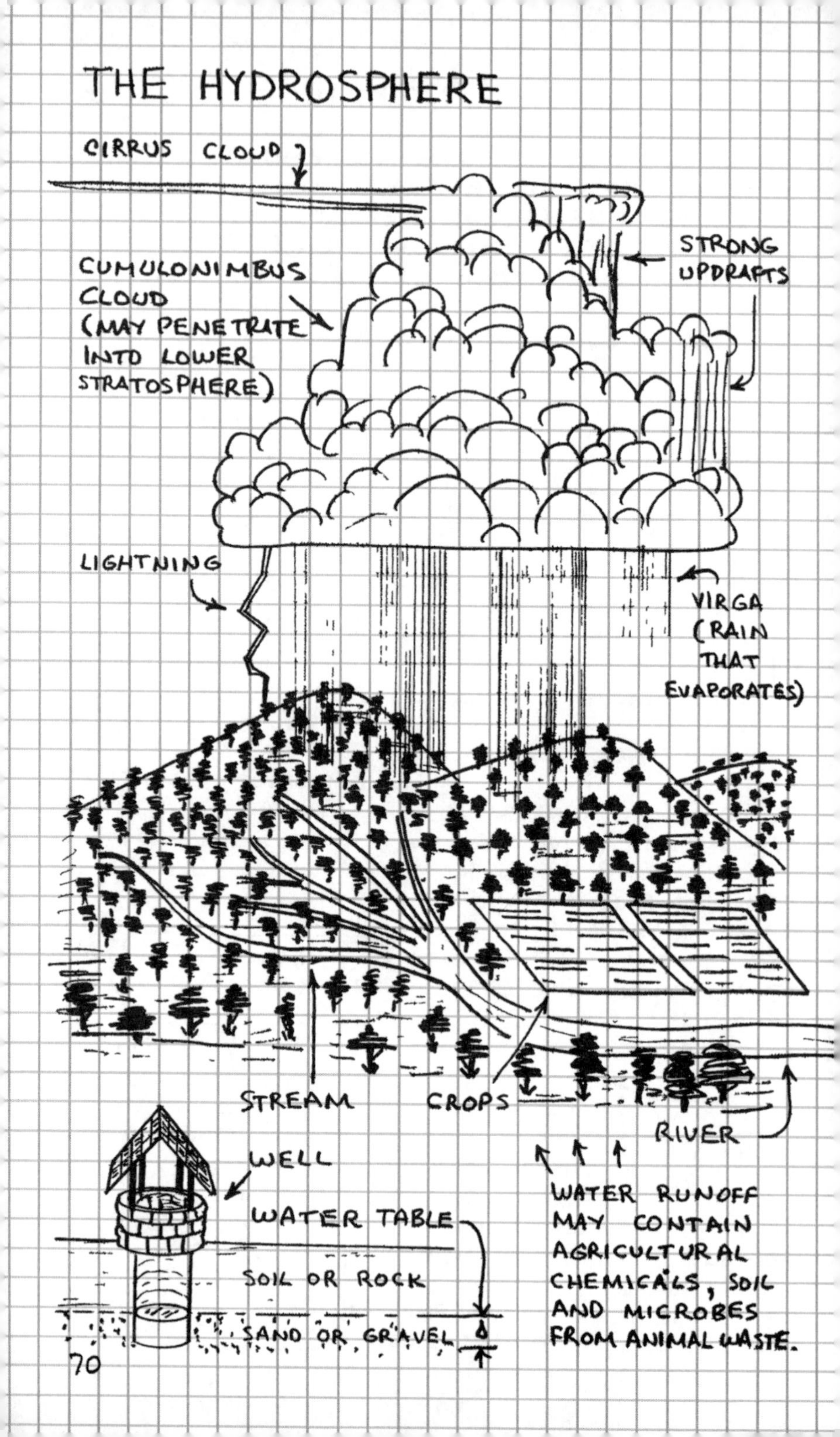
THE HYDROSPHERE
CIRRUS CLOUD
CUMULONIMBUS CLOUD (MAY PENETRATE INTO LOWER STRATOSPHERE)
STRONG UPDRAFTS
LIGHTNING
VIRGA (RAIN THAT EVAPORATES)
STREAM
CROPS
RIVER
WELL
WATER TABLE
SOIL OR ROCK
SAND OR GRAVEL
WATER RUNOFF MAY CONTAIN AGRICULTURAL CHEMICALS, SOIL AND MICROBES FROM ANIMAL WASTE.

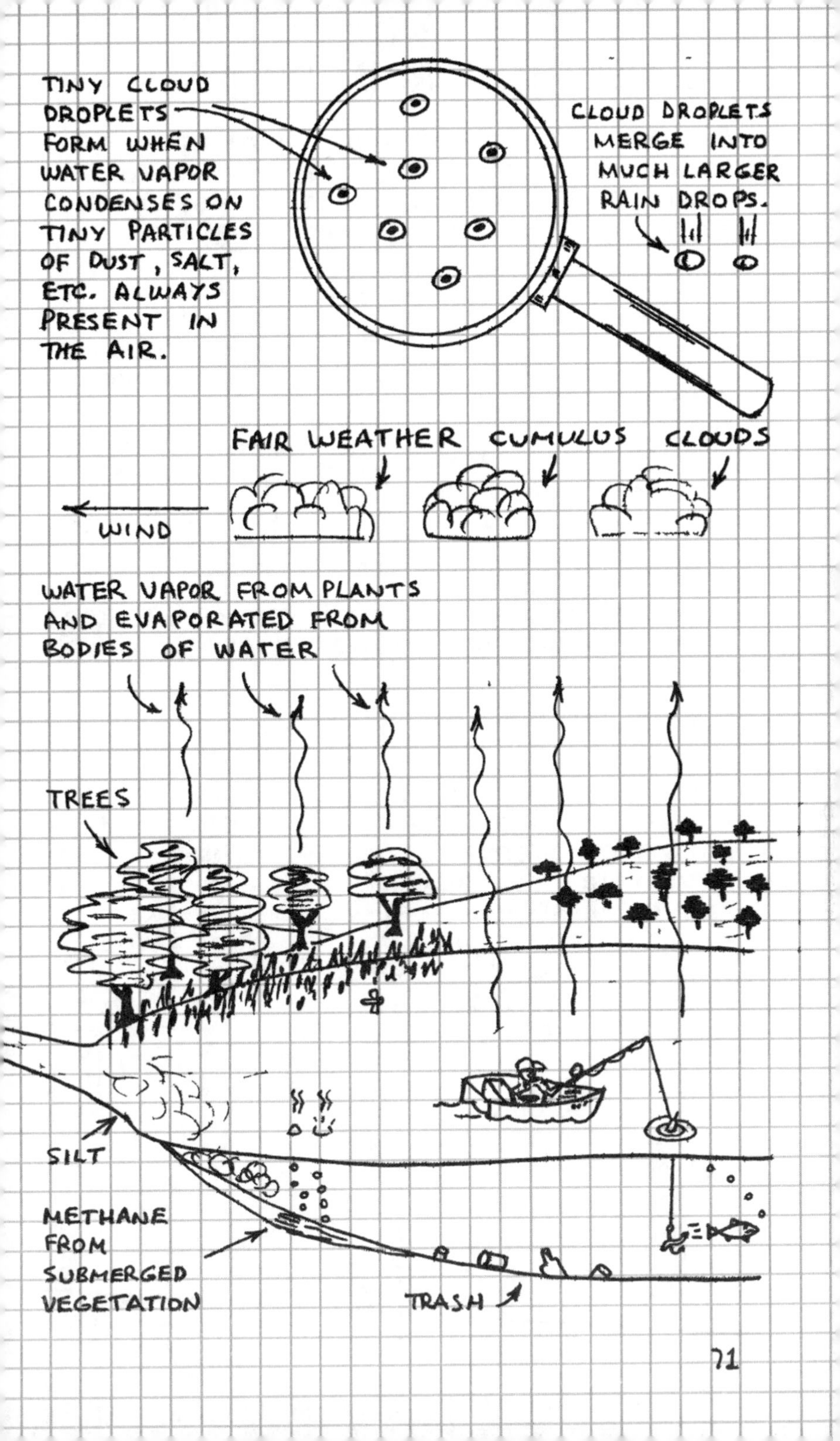
TINY CLOUD DROPLETS FORM WHEN WATER VAPOR CONDENSES ON TINY PARTICLES OF DUST, SALT, ETC. ALWAYS PRESENT IN THE AIR.
CLOUD DROPLETS MERGE INTO MUCH LARGER RAIN DROPS.
FAIR WEATHER CUMULUS CLOUDS
WIND
WATER VAPOR FROM PLANTS AND EVAPORATED FROM BODIES OF WATER
TREES
SILT
METHANE FROM SUBMERGED VEGETATION
TRASH

MEASURING RAIN DROPS

A TYPICAL RAIN DROP HAS A DIAMETER OF ABOUT 2 MILLIMETERS (O). DROPS CAN BE SMALLER OR LARGER. USE THIS INSTRUMENT TO STUDY THE SIZE OF RAIN DROPS:

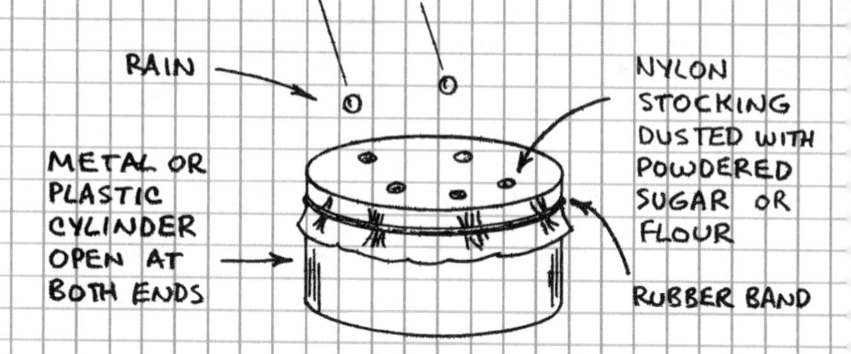

MEASURE AND RECORD DIAMETER OF INDIVIDUAL DROPS AND AVERAGE DIAMETER OF ALL DROPS.

RAIN DROP DETECTOR

THIS SIMPLE CIRCUIT WILL SOUND A TONE WHEN A RAIN DROP SPLASHES ON A SENSOR. THE SENSOR CAN BE ALUMINUM SCREEN MOUNTED JUST ABOVE A METAL PLATE (e.g. COPPER FOIL ON PC BOARD). OR MAKE A "COMB" SENSOR:

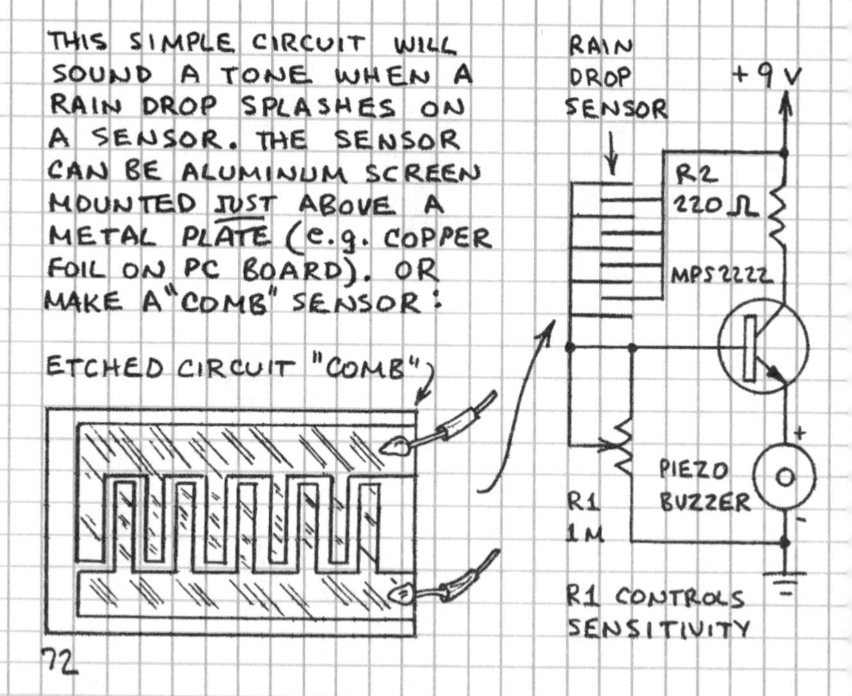

MEASURING RAIN FALL

MEASURING THE AMOUNT OF RAIN AND SNOW IS AN IMPORTANT PART OF ENVIRONMENTAL MONITORING.

USE STORE-BOUGHT RAIN GAUGE OR MAKE YOUR OWN USING A CLEAR PLASTIC CYLINDER WITH A FLAT BOTTOM. PLACE GAUGE AWAY FROM TREES AND BUILDINGS. NOTE: WIND MAY REDUCE THE RAIN COLLECTED BY THE GAUGE BY UP TO 10 %.

ADD A FUNNEL TO INCREASE ACCURACY WHEN MEASURING SMALL AMOUNTS OF RAIN. DIVIDE AREA OF LARGE END OF FUNNEL BY AREA OF INSIDE, OPEN END OF GAUGE TO GET CORRECTION FACTOR. DIVIDE HEIGHT OF WATER IN GAUGE BY CORRECTION FACTOR TO GET ACTUAL RAIN FALL.

NUMBER OF RAIN DROPS

THE VOLUME OF A SPHERICAL RAIN DROP WITH A DIAMETER OF 2.5mm IS 8.18 mm^3 ($V = \frac{4}{3}\pi r^3$).

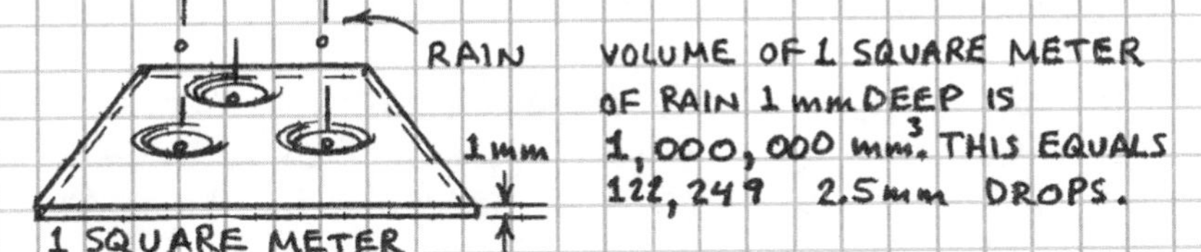

VOLUME OF 1 SQUARE METER OF RAIN 1 mm DEEP IS 1,000,000 mm^3. THIS EQUALS 122,249 2.5mm DROPS.

122,249 DROPS PER MILLIMETER OF RAIN PER SQUARE METER IS 122,249,000,000 DROPS PER SQUARE KILOMETER OR 316,623,456,459 DROPS PER SQUARE MILE !

DEW

DEW IS LIQUID WATER THAT CONDENSES ON COOL OBJECTS. UP TO 0.6 mm (0.02 INCH) OF DEW MAY CONDENSE ON EXPOSED OBJECTS AND PLANTS AT NIGHT.

DEW POINT

THE TEMPERATURE AT WHICH DEW BEGINS TO FORM IS THE DEW POINT. THE DEW POINT REVEALS MUCH ABOUT THE LOCAL WEATHER:

1. THE TEMPERATURE AT NIGHT USUALLY DOES NOT FALL BELOW THE DEW POINT.

2. A DEW POINT OF 20°C (68°F) OR HIGHER AND AN APPROACHING COLD FRONT MEANS THUNDERSTORMS ARE POSSIBLE.

3. EXPECT FOG IF THE PREDICTED LOW TEMPERATURE MATCHES THE DEW POINT.

4. A DEW POINT OF 20°C (68°F) OR HIGHER MEANS THE AIR IS UNCOMFORTABLY HUMID.

5. WHEN THE DEW POINT IS BELOW FREEZING, FROST MAY FORM ON EXPOSED SURFACES.

6. DEW THAT FREEZES FORMS A GLAZE OF ICE.

MEASURING DEW POINT

IF YOU MAKE A WET/DRY RELATIVE HUMIDITY INSTRUMENT (HYGROMETER, SEE P15), THE DEW POINT IS APPROXIMATELY:

$$D.P. = (5T_{WET} - 2T_{DRY})/3 \quad (°CELSIUS)$$

T IS TEMPERATURE. FORMULA FROM "CLIMATE DATA AND RESOURCES" BY E. LINACRE (ROUTLEDGE, 1992).

PRECIPITABLE WATER

TOP OF ATMOSPHERE

EARTH

CONDENSING THE WATER VAPOR IN A COLUMN THROUGH THE ATMOSPHERE YIELDS THE PRECIPITABLE WATER. C.H. REITAN* HAS DEVISED A FORMULA THAT ESTIMATES THE PRECIPITABLE WATER:

$$\ln W = (0.061 \times D.P.) - 0.11$$

ln W IS NATURAL LOG OF PRECIPITABLE WATER (cm). D.P. IS DEW POINT (°C).

*JOURNAL OF APPLIED METEOROLOGY, VOL. 2, 776-9, 1963.

CLOUD HEIGHT

CUMULUS CLOUDS FORM WHEN WARM, HUMID AIR RISES TO WHERE THE AIR TEMPERATURE FALLS BELOW THE DEW POINT. KNOWING THAT AIR TEMPERATURE FALLS ABOUT 2.77°C (5.5°F) PER 0.3 KILOMETER (1,000 FEET), LESLIE TROWBRIDGE* DERIVED THIS FORMULA FOR ESTIMATING THE HEIGHT OF THE BASE OF A CUMULUS CLOUD:

HEIGHT (FEET) = 227 × (T − D.P.)

T = GROUND TEMPERATURE (°F)
D.P. = DEW POINT (°F)

SUNLIGHT WARMS THE EARTH. COLUMNS OF WARM, MOIST AIR THEN RISE IN THE SKY.

DESCENDING, COOL AIR

*"EXPERIMENTS IN METEOROLOGY," DOUBLEDAY, p. 239, 1974.

MEASURING CLOUDINESS

THE TEMPERATURE OF EARTH IS REGULATED IN PART BY CLOUDS. WARM AIR CAN CONTAIN MORE WATER VAPOR, HENCE MORE CLOUDS. THE CLOUDS REFLECT SUNLIGHT BACK INTO SPACE, THUS COOLING THE EARTH. RECORDING THE FRACTION OF THE SKY COVERED BY CLOUDS CAN PROVIDE IMPORTANT INFORMATION ABOUT THE EFFECT OF CLOUDS ON CLIMATE. THE FRACTION OF SKY COVERED BY CLOUDS IS MEASURED IN TENTHS OR OCTAS (EIGHTHS):

0 TENTHS OR 0 OCTAS = CLOUD-FREE SKY
5 TENTHS OR 4 OCTAS = 50% CLOUDINESS
10 TENTHS OR 8 OCTAS = OVERCAST SKY

ESTIMATE CLOUDINESS IN EACH QUADRANT OF THE COMPASS. AVERAGE THE 4 ESTIMATES TO GET THE OVERALL CLOUDINESS.

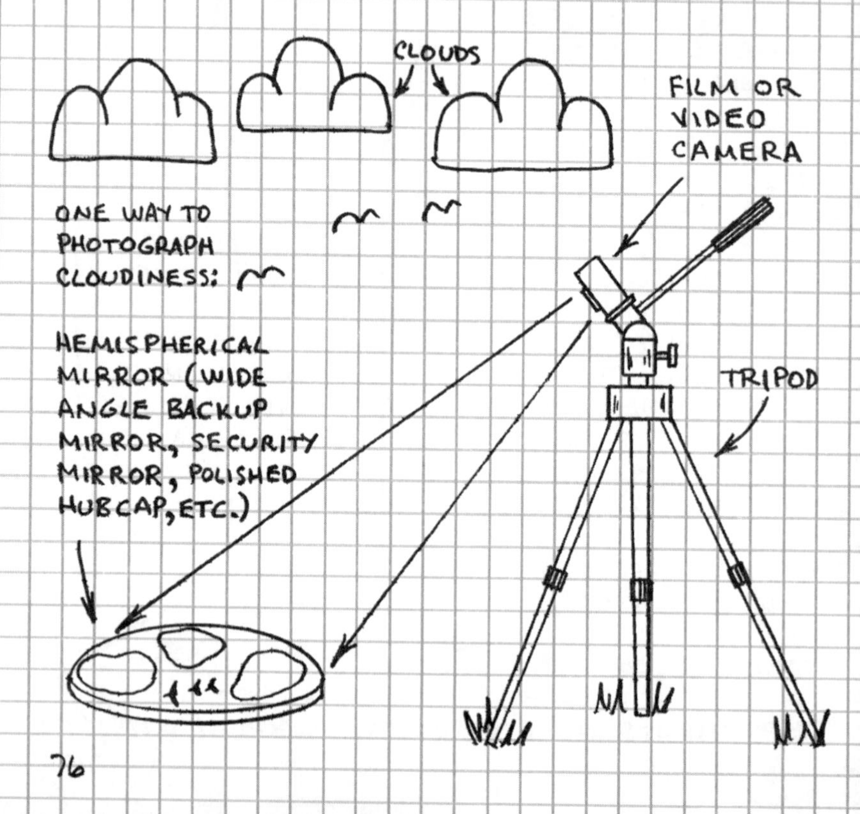

STUDYING LIGHTNING

THE AIR IN THE PATH OF A LIGHTNING BOLT IS HEATED ALMOST INSTANTLY TO 30,000°C (54,000°F). THE PRESSURE OF THIS AIR CAN BE 10 TO 100 TIMES THE PRESSURE AT SEA LEVEL. THE RESULTING SHOCK WAVES CAUSES THE SOUND HEARD AS THUNDER.

YOU CAN USE A DIGITAL STOP WATCH TO MEASURE DISTANCE BETWEEN YOU AND A LIGHTNING BOLT AND TO ESTIMATE THE TOTAL LENGTH OF LIGHTNING BOLTS.

CAUTION: STAY INDOORS WHEN CONDUCTING THESE EXPERIMENTS!

DISTANCE TO BOLT: START STOPWATCH WHEN YOU SEE FLASH AND STOP WHEN YOU HEAR THUNDER. DISTANCE IS ELAPSED SECONDS TIMES 1125 (FEET) OR 343 (METERS).

LENGTH OF BOLT: START STOPWATCH WHEN YOU FIRST HEAR THUNDER AND STOP WHEN THUNDER ENDS. MINIMUM LENGTH OF THE BOLT IS ELAPSED SECONDS TIMES 1.86 (MILES) OR 3 (KILOMETERS). SEE "THUNDER" BY A. FEW, *SCIENTIFIC AMERICAN*, JULY 1975.

LIGHTNING STRUCK ELM TREE NEAR MY BARN, SPLITTING TRUNK IN HALF, BREAKING LARGE BRANCH AND BLOWING BARK AND LARGE SPLINTERS FROM TRUNK.

WATER TURBIDITY

SUSPENDED PARTICLES, LIQUID CONTAMINANTS AND WATER MOLECULES ALL ABSORB OR SCATTER LIGHT PASSING THROUGH WATER. THE SECCHI DISK PROVIDES A SIMPLE, TIME-TESTED MEANS FOR MEASURING WATER CLARITY.

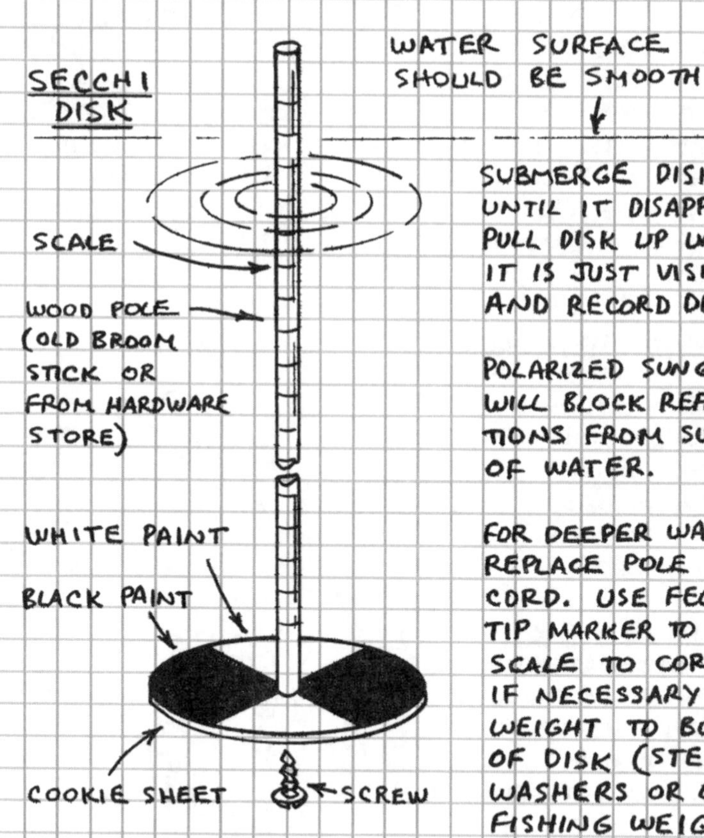

SUBMERGE DISK UNTIL IT DISAPPEARS. PULL DISK UP UNTIL IT IS JUST VISIBLE AND RECORD DEPTH.

POLARIZED SUNGLASSES WILL BLOCK REFLECTIONS FROM SURFACE OF WATER.

FOR DEEPER WATER REPLACE POLE WITH CORD. USE FELT TIP MARKER TO ADD SCALE TO CORD. IF NECESSARY, ADD WEIGHT TO BOTTOM OF DISK (STEEL WASHERS OR LEAD FISHING WEIGHTS).

<u>WATER</u> <u>VISIBILITY</u> <u>RECORD</u>

ON JUNE 27, 1676, ON THE SEA EAST OF NOVAYA ZEMLYA, CAPTAIN JOHN WOOD OBSERVED SHELLS ON THE BOTTOM "IN 80 FATHOMS WATER, WHICH IS 480 FEET...." (EOS, MARCH 1, 1994, P. 99.)

USE CAUTION WITH SECCHI DISK!

ELECTRONIC TURBIDIMETER

THIS CIRCUIT MEASURES THE CLARITY OF A LIQUID WITH RESPECT TO THAT OF CLEAR WATER.

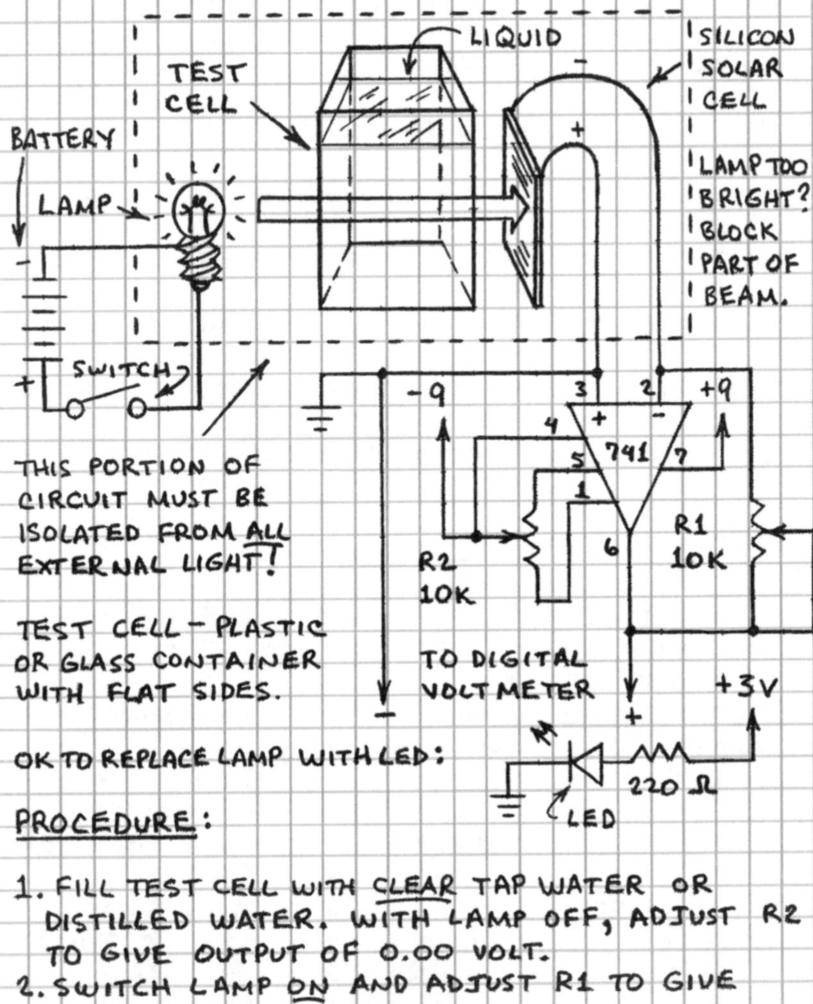

OK TO REPLACE LAMP WITH LED:

PROCEDURE:

1. FILL TEST CELL WITH CLEAR TAP WATER OR DISTILLED WATER. WITH LAMP OFF, ADJUST R2 TO GIVE OUTPUT OF 0.00 VOLT.
2. SWITCH LAMP ON AND ADJUST R1 TO GIVE OUTPUT OF 1.00 VOLT.
3. INSERT CELL WITH SAMPLE WATER AND RECORD OUTPUT VOLTAGE.

FOR MORE SENSITIVITY, INCREASE ±9 VOLTS TO ±12 VOLTS AND ADJUST R1 TO GIVE 8-10 VOLTS OUT WITH CLEAN WATER IN TEST CELL.

TESTING WATER

WATER IS SOMETIMES DESCRIBED AS THE UNIVERSAL SOLVENT. YOU CAN EASILY MEASURE THE CONCENTRATION OF VARIOUS IMPURITIES IN WATER USING TEST KITS FROM AQUARIUM SUPPLY STORES AND RADIO SHACK.

<u>IMPORTANT</u> <u>TESTS</u>:

HARDNESS – CAUSED BY DISSOLVED MINERALS.

AMMONIA – A WASTE PRODUCT OF BACTERIA.

NITRATE – INGREDIENT OF CROP FERTILIZER.

NITRITE – IMPAIRS ABILITY OF BLOOD TO CARRY OXYGEN.

CHLORINE – DISINFECTANT OFTEN ADDED TO WATER.

PH – CONCENTRATION OF HYDROGEN IONS.

THE pH SCALE

AN INCREASE OF 1 pH IS AN INCREASE OF 10 TIMES THE NUMBER OF HYDROGEN IONS

	pH	
↑ HIGH (ALKALINE)	14	LYE
	13	BLEACH
	12	
	11	AMMONIA
	10	
	9	BAKING SODA
	8	SEA WATER
NEUTRAL	7	DISTILLED WATER
	6	MILK
	5	MANY FOODS
LOW (ACID) ↓	4	ORANGE JUICE
	3	VINEGAR
	2	LEMON JUICE
	1	
	0	BATTERY ACID

RAIN WATER FALLING THROUGH UNPOLLUTED AIR HAS A pH OF ABOUT 5.6.

DO-IT-YOURSELF PH INDICATOR – LIQUIFY PURPLE CABBAGE IN A BLENDER. THE PURPLE JUICE WILL CHANGE COLOR AS PH CHANGES. DILUTE TO USE.

WATER AND CARBON DIOXIDE

WATER READILY ABSORBS CARBON DIOXIDE (CO_2), WHICH MAKES POSSIBLE CARBONATED BEVERAGES. MUCH OF THE CO_2 IN THE AIR IS ABSORBED BY THE OCEAN. RAIN ABSORBS CO_2, WHICH FORMS CARBONIC ACID AND CAUSES RAIN FALLING THROUGH CLEAN AIR TO BE MILDLY ACIDIC.

TO DEMONSTRATE ABSORPTION OF CO_2 IN WATER, BLOW BUBBLES THROUGH SMALL CUP OF WATER. USE pH INDICATOR DROPS OR PAPER TO MEASURE pH OF THE WATER BEFORE AND AFTER BLOWING. I MEASURED CHANGE IN pH FROM 6.2 TO 6.0 AFTER BLOWING 2 MINUTES INTO TEST TUBE.

ACTIVATED CARBON AND WATER

ACTIVATED CARBON IS A HIGHLY POROUS FORM OF CHARCOAL. IT IS WIDELY USED TO REMOVE IMPURITIES FROM DRINKING WATER AND WATER IN AQUARIUMS. ACTIVATED CARBON IS SOLD BY AQUARIUM STORES. THIS SIMPLE DEMONSTRATION SHOWS ITS ABILITY TO FILTER WATER.

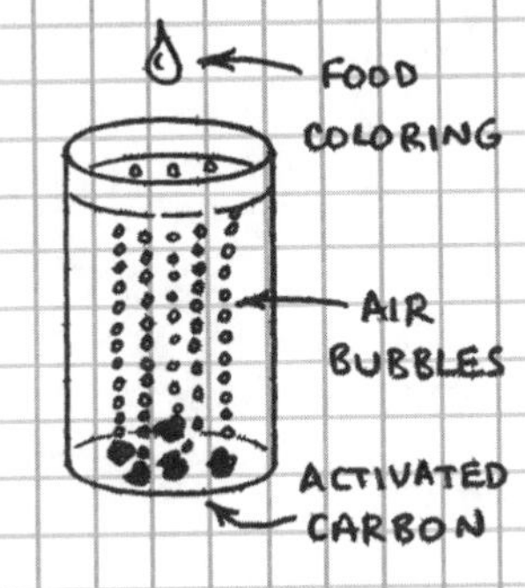

ADD ACTIVATED CARBON TO WATER AND IT WILL FIZZ AS TINY AIR BUBBLES ARE RELEASED. ADD A DROP OF FOOD COLORING, CAP THE CONTAINER AND SHAKE VIGOROUSLY. WATER WILL BECOME CLEAR AS CARBON ABSORBS THE COLORED DYE.

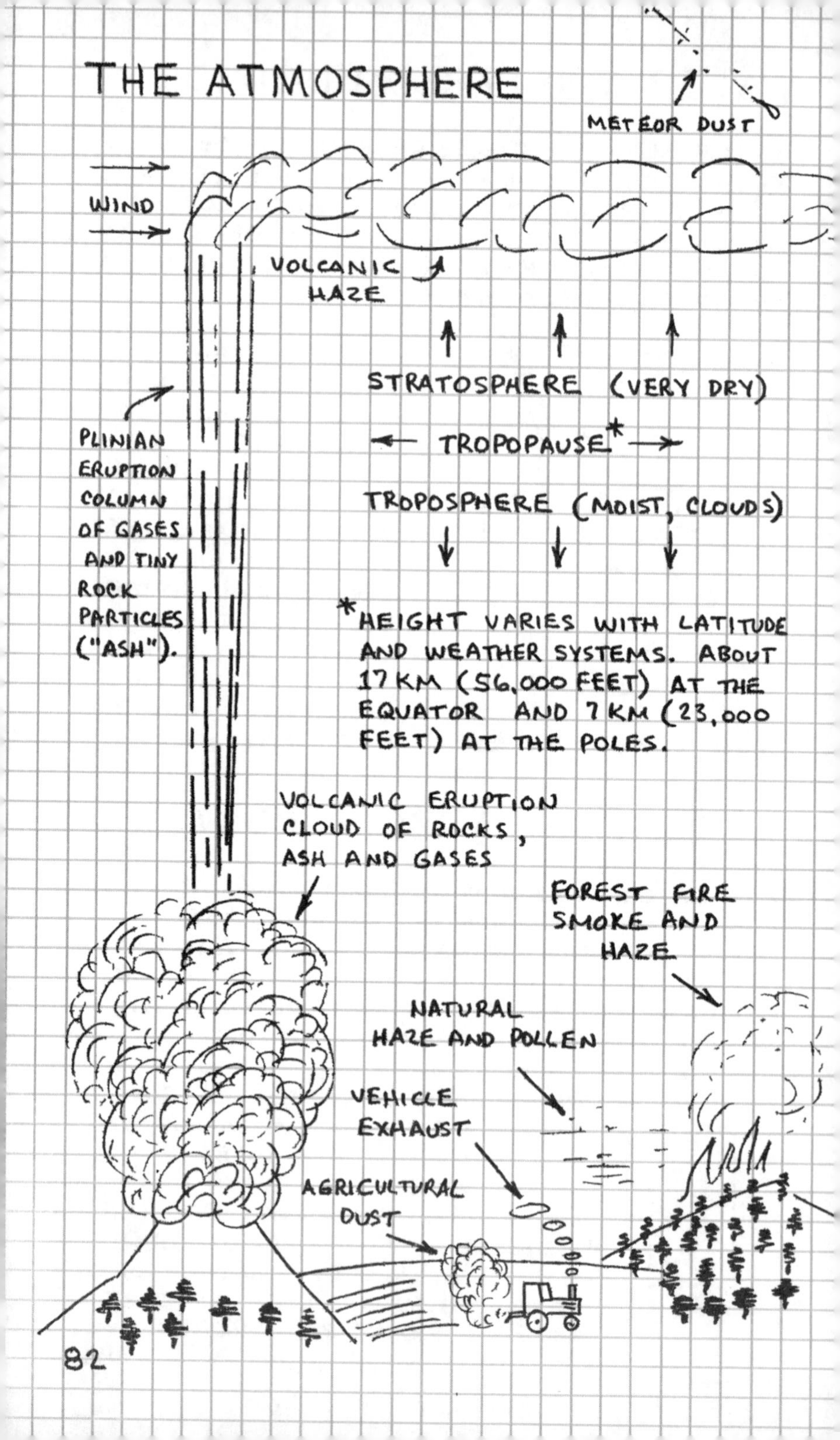
THE ATMOSPHERE
METEOR DUST
WIND
VOLCANIC HAZE
STRATOSPHERE (VERY DRY)
TROPOPAUSE*
TROPOSPHERE (MOIST, CLOUDS)
PLINIAN ERUPTION COLUMN OF GASES AND TINY ROCK PARTICLES ("ASH").
*HEIGHT VARIES WITH LATITUDE AND WEATHER SYSTEMS. ABOUT 17 KM (56,000 FEET) AT THE EQUATOR AND 7 KM (23,000 FEET) AT THE POLES.
VOLCANIC ERUPTION CLOUD OF ROCKS, ASH AND GASES
FOREST FIRE SMOKE AND HAZE
NATURAL HAZE AND POLLEN
VEHICLE EXHAUST
AGRICULTURAL DUST

THE OZONE LAYER INCLUDES ABOUT 90% OF THE TOTAL OZONE. THE REMAINDER IS IN THE TROPOSPHERE.

SUN

ULTRAVIOLET RAYS

OZONE LAYER ABSORBS MOST OF THE SUN'S ULTRAVIOLET RADIATION. VOLCANIC HAZE AND BOTH NATURAL AND ANTHROPOGENIC* GASES CAN DESTROY OZONE. *MAN MADE

OZONE LAYER IS 15-35 KM (49-115,000 FEET).

<u>ATMOSPHERIC CONTENTS</u>: 78% NITROGEN, 21% OXYGEN, 1% ARGON PLUS A DOSE OF WATER VAPOR, OZONE, CARBON DIOXIDE, METHANE, CARBON MONOXIDE, SULFUR DIOXIDE, SMOKE, DUST, SPIDER WEBS, POLLEN, INSECTS, BACTERIA AND DOZENS OF OTHER GASES AND PARTICLES.

CONDENSATION TRAIL (CONTRAIL)

CUMULUS CLOUD

HAZE CAUSED BY SULFATE PARTICLES COATED WITH WATER

SULFUR DIOXIDE GAS

URBAN HAZE

ACIDIFIED RAIN

CITY

POWER PLANT

TRAFFIC

COAL

THE SOLAR RADIATION BUDGET

THE OZONE LAYER ABSORBS MOST ULTRAVIOLET AND SOME ORANGE LIGHT.

SUN

VOLCANIC HAZE

REFLECTED BACK TO SPACE

CIRRUS CLOUD

DIFFUSE RADIATION (SCATTERED)

DIRECT SUN RADIATION

CUMULUS CLOUD

SMOKE

DIFFUSE RADIATION

RADIATION SCATTERED FROM SIDE OF CUMULUS CLOUD.*

CLOUD SHADOW

TOTAL RADIATION IS DIRECT SUN + DIFFUSE RADIATION ON A FLAT SURFACE.

* CUMULUS CLOUDS IN CLEAR SKY CAN INCREASE SOLAR ULTRAVIOLET BY UP TO 25 %. (SEE F. MIMS III & J. FREDERICK <u>NATURE</u> 371, p. 291, 1994.)

HAZE AND SOLAR RADIATION

NATURAL HAZE IS CAUSED BY SMOKE FROM FOREST FIRES, WATER VAPOR FOG, VERY THIN OVERCAST CIRRUS OR STRATUS CLOUDS, DUST, SEA SALT AND PHOTOCHEMICAL REACTIONS OF SUNLIGHT AND VARIOUS GASES EMITTED BY PLANTS.

ANTHROPOGENIC HAZE, A BYPRODUCT OF HUMAN ACTIVITY, IS CAUSED BY EMISSIONS FROM COAL-FIRED POWER PLANTS, FIREPLACE SMOKE, CONTRAILS FROM HIGH ALTITUDE AIRCRAFT (WHICH CAN COVER MUCH OF THE SKY) AND PHOTOCHEMICAL REACTIONS OF SUNLIGHT AND GASES EMITTED BY INTERNAL COMBUSTION ENGINES. ANTHROPOGENIC HAZE IS ESPECIALLY BAD OVER THE EASTERN PORTIONS OF EUROPE AND THE UNITED STATES.

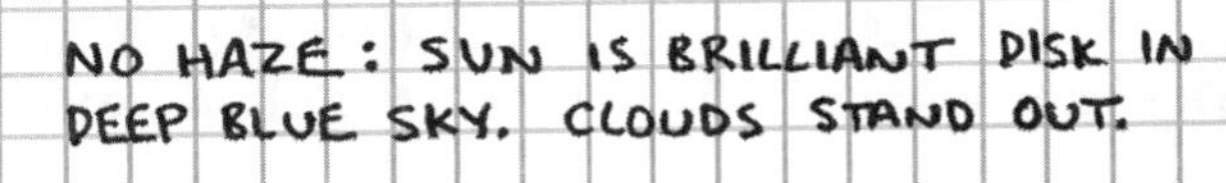

NO HAZE: SUN IS BRILLIANT DISK IN DEEP BLUE SKY. CLOUDS STAND OUT.

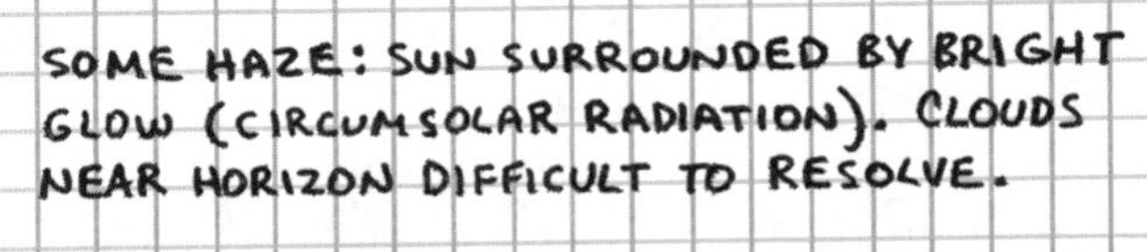

SOME HAZE: SUN SURROUNDED BY BRIGHT GLOW (CIRCUMSOLAR RADIATION). CLOUDS NEAR HORIZON DIFFICULT TO RESOLVE.

CONSIDERABLE HAZE: SUN DIM. ENTIRE SKY PALE, MILKY BLUE. CLOUDS BLEND INTO HAZE AND DIFFICULT TO SEE.

HAZE SIGNIFICANTLY REDUCES DIRECT RADIATION, SIGNIFICANTLY INCREASES DIFFUSE RADIATION AND SLIGHTLY REDUCES TOTAL RADIATION.

HAZE SCATTERS SOME RADIATION BACK INTO SPACE, THUS CAUSING A COOLING EFFECT.

HAZE GREATLY INCREASES DIFFUSE RADIATION ON PLANTS AND ANIMALS SHADED FROM DIRECT SUN. DURING SUMMER OF 1994, I FOUND THAT A PERSON SHADED FROM DIRECT SUN BY A SMALL UMBRELLA CAN RECEIVE 30% OR MORE SOLAR ULTRAVIOLET AT A HAZY SITE NEAR SEA LEVEL THAN AT THE TOP OF PIKES PEAK (ELEVATION: 4,301 METERS OR 14,110 FEET).

ATMOSPHERIC OPTICAL THICKNESS

ATMOSPHERIC OPTICAL THICKNESS (AOT) IS A MEASURE OF THE CLARITY OF THE AIR IN A VERTICAL COLUMN THROUGH THE ATMOSPHERE. AOT INDICATES THE AMOUNT OF HAZE, SMOG, SMOKE, DUST AND VOLCANIC AEROSOLS IN THE ATMOSPHERE. A SMALL AOT INDICATES A CLEAN ATMOSPHERE.

YOU CAN MEASURE AOT WITH A SUN PHOTOMETER LIKE THE ONE ON PAGE 39 AND A CALCULATOR WITH A $\ln$ (NATURAL LOGARITHM) KEY. A SIMPLIFIED FORMULA FOR AOT IS:

$$AOT = (\ln I_0 / \ln I) / m$$

I_0 IS THE SIGNAL THE SUN PHOTOMETER WOULD MEASURE ABOVE THE ATMOSPHERE— THE EXTRATERRESTRIAL (ET) CONSTANT.

I IS THE SIGNAL DURING A SPECIFIC SUN OBSERVATION.

m IS THE AIR MASS (SEE BELOW) DURING THE OBSERVATION.

AIR MASS (m)

AIR MASS IS $1/\sin\theta$, WHERE θ IS THE ANGLE OF THE SUN ABOVE THE HORIZON.

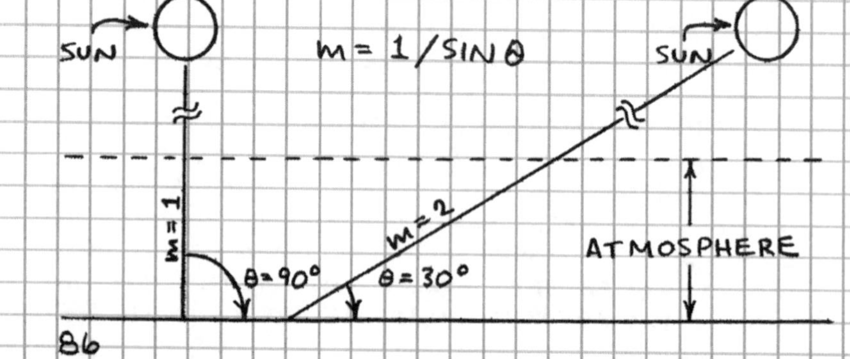

LED SUN PHOTOMETER

LIGHT-EMITTING DIODES EMIT AND DETECT LIGHT OVER A RELATIVELY NARROW BAND OF WAVELENGTHS (30-150 NANOMETERS). THIS MEANS AN LED CAN BE USED IN A SUN PHOTOMETER WITHOUT AN EXTERNAL OPTICAL FILTER.

REFERENCE: F.M. MIMS III, "SUN PHOTOMETER WITH LEDs..." APPLIED OPTICS VOL 31, NO 33, 6965-6967, 1992.

SOLAR CELL RADIOMETER

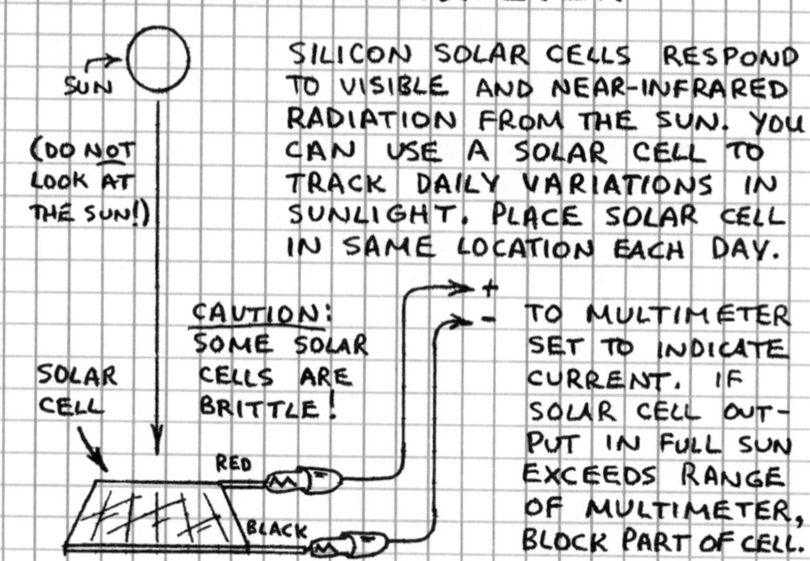

SOLAR CELL SUN PHOTOMETER

THE OPTICAL THICKNESS OF THE ATMOSPHERE (SEE PAGE 86) CAN BE MEASURED WITH A RADIOMETER THAT RESPONDS TO A NARROW BAND OF LIGHT WAVELENGTHS. A SOLAR CELL RADIOMETER CAN BE CONVERTED INTO A SUN PHOTOMETER.

SUN

COLLIMATOR TUBE TO RESTRICT FIELD OF VIEW TO THE SUN

BLACK PAINT OR PAPER

FILTER*

SOLAR CELL

TO MULTIMETER

* USE PLASTIC OR GLASS CAMERA FILTER OR COLORED PLASTIC INDEX TAB.

GRAPH YOUR DATA

HOW TO MEASURE THE SUN'S ANGLE

USE VARIOUS ASTRONOMY COMPUTER PROGRAMS TO FIND THE SUN'S ANGLE. OR MEASURE THE SUN'S ANGLE TO WITHIN ABOUT ± 1° WITH A SUN ANGLE INDICATOR:

SUN

BUBBLE LEVEL (HARDWARE STORE)

"L" BRACKET

SCALE

θ a b

$\tan \theta = \frac{a}{b}$

HOW TO MEASURE THE ET* CONSTANT

FIRST, MEASURE I FOR 1/2 DAY, EVERY 30 MINUTES NEAR NOON, MORE OFTEN AT LOWER SUN ANGLES. THEN PLOT $\ln$ OF I VS. m AT EACH SUN OBSERVATION. DRAW A STRAIGHT LINE THROUGH THE POINTS. THE $\ln$ OF THE ET CONSTANT IS WHERE THE LINE INTERCEPTS THE VERTICAL (Y) AXIS WHERE m = 0. <u>HINT</u>: USE THE LINEAR REGRESSION FEATURE OF A CALCULATOR OR COMPUTER SPREADSHEET TO FIND THE INTERCEPT AT m = 0. *EXTRATERRESTRIAL

I_0 (THE INTERCEPT AT M = 0) SHOULD AGREE WITHIN 5%.

CLEAR DAY

WHEN SKY IS CLEAR, AOT IS LOWER IN WINTER THAN IN SUMMER.

HAZY DAY

$\ln$ INTENSITY (I)

0 1 2 3 4 5 6

AIR MASS (m)

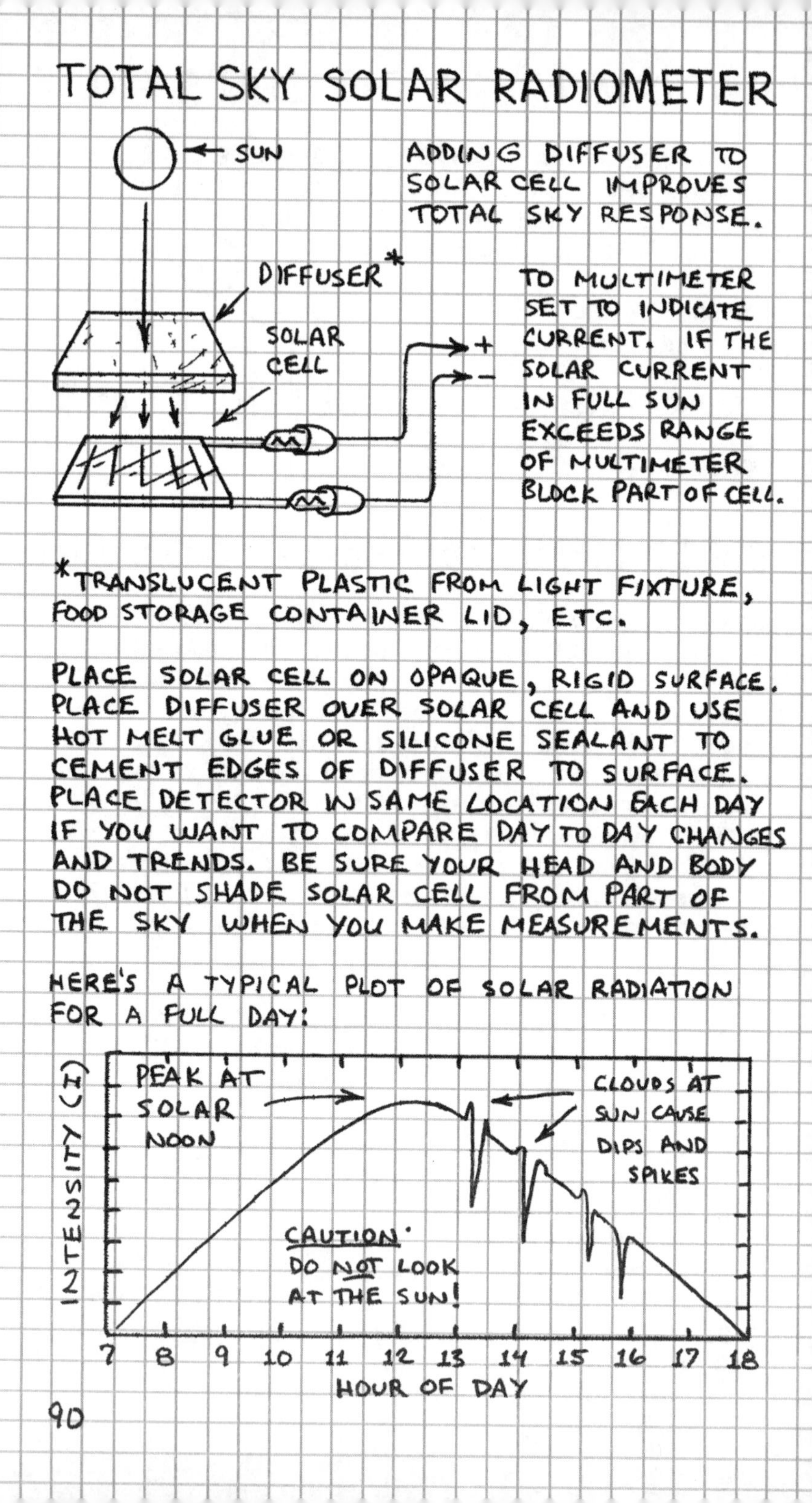

TOTAL SKY SOLAR RADIOMETER

ADDING DIFFUSER TO SOLAR CELL IMPROVES TOTAL SKY RESPONSE.

TO MULTIMETER SET TO INDICATE CURRENT. IF THE SOLAR CURRENT IN FULL SUN EXCEEDS RANGE OF MULTIMETER BLOCK PART OF CELL.

*TRANSLUCENT PLASTIC FROM LIGHT FIXTURE, FOOD STORAGE CONTAINER LID, ETC.

PLACE SOLAR CELL ON OPAQUE, RIGID SURFACE. PLACE DIFFUSER OVER SOLAR CELL AND USE HOT MELT GLUE OR SILICONE SEALANT TO CEMENT EDGES OF DIFFUSER TO SURFACE. PLACE DETECTOR IN SAME LOCATION EACH DAY IF YOU WANT TO COMPARE DAY TO DAY CHANGES AND TRENDS. BE SURE YOUR HEAD AND BODY DO NOT SHADE SOLAR CELL FROM PART OF THE SKY WHEN YOU MAKE MEASUREMENTS.

HERE'S A TYPICAL PLOT OF SOLAR RADIATION FOR A FULL DAY:

SHADOW BAND RADIOMETER

A SHADOW BAND (OR RING) IS A STRIP OF OPAQUE, FLEXIBLE PLASTIC, METAL OR STIFF PAPER BENT INTO A HALF CIRCLE. THE BAND IS ORIENTED EAST AND WEST AND TILTED TO FACE THE SUN. A LIGHT SENSOR UNDER THE BAND WILL BE SHADED AS THE SUN MOVES ACROSS THE SKY. IT WILL THEN RECEIVE ONLY THE DIFFUSE RADIATION FROM THE SKY AND CLOUDS.

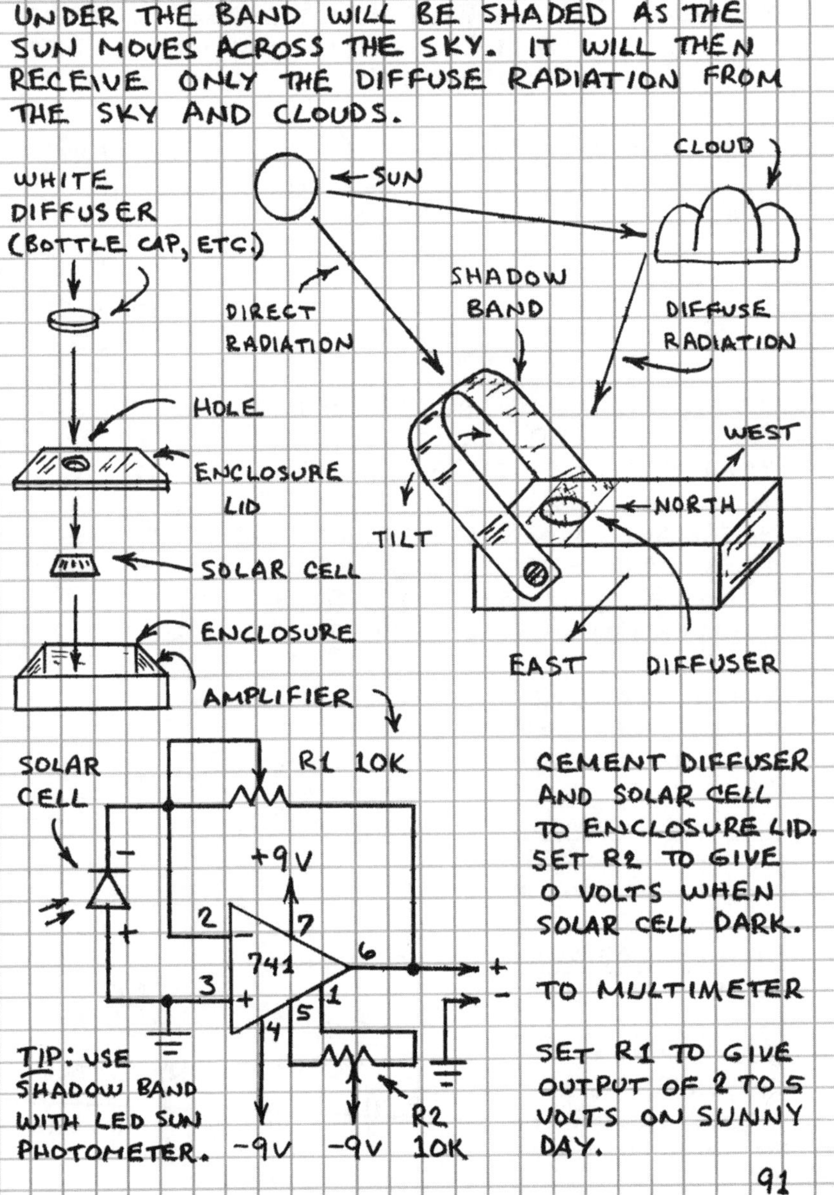

CEMENT DIFFUSER AND SOLAR CELL TO ENCLOSURE LID. SET R2 TO GIVE 0 VOLTS WHEN SOLAR CELL DARK.

SET R1 TO GIVE OUTPUT OF 2 TO 5 VOLTS ON SUNNY DAY.

TIP: USE SHADOW BAND WITH LED SUN PHOTOMETER.

MEASURING TOTAL AND DIFFUSE RADIATION

WHEN THE ATMOSPHERIC OPTICAL THICKNESS (AOT) MEASURED BY A SUN PHOTOMETER IS HIGH, THE DIRECT SOLAR RADIATION IS REDUCED AND DIFFUSE RADIATION IS INCREASED. THE LED SUN PHOTOMETER ON PAGE 39 CAN BE MODIFIED TO MEASURE THE TOTAL AND DIFFUSE RADIATION AND THE RATIO OF THE DIFFUSE OR DIRECT TO THE TOTAL RADIATION. FIRST MODIFY THE LED LIKE THIS:

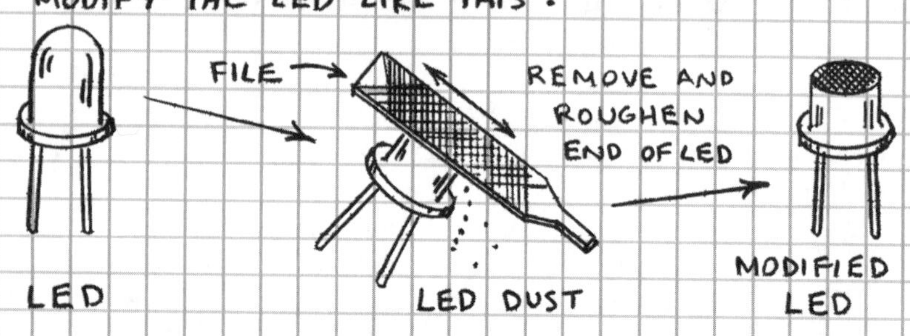

NEXT, ARRANGE THE PHOTOMETER SO THE FLAT TOP OF THE LED LOOKS STRAIGHT UP AT THE ZENITH SKY. USE A BUBBLE LEVEL TO MAKE SURE THE PHOTOMETER IS LEVEL EACH TIME YOU MAKE A MEASUREMENT. ADJUST THE RESISTANCE OF R1 FOR BEST RESULTS — BUT MAKE ANY CHANGE PERMANENT SO YOUR MEASUREMENTS WILL BE COMPARABLE.

<u>TOTAL RADIATION</u>
OUTPUT WHEN LED POINTED AT ZENITH.

<u>DIFFUSE RADIATION</u>
OUTPUT WHEN LED SHADED AS SHOWN.

<u>DIRECT RADIATION</u>
TOTAL − DIFFUSE

SUGGESTION:
TRACK RATIO OF DIRECT OR DIFFUSE TO TOTAL OVER TIME.

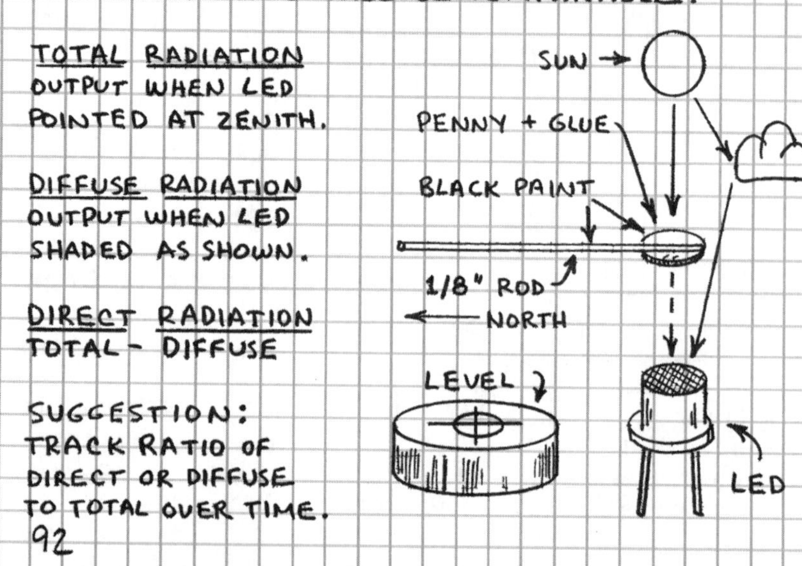

MEASURING THE SOLAR AUREOLE

THE RING OF LIGHT AROUND THE SUN ON ALL BUT THE CLEAREST DAYS IS THE SOLAR AUREOLE OR THE CIRCUMSOLAR RADIATION. THE SIZE AND BRIGHTNESS OF THE AUREOLE IS DETERMINED BY HAZE. YOU CAN USE A SUN PHOTOMETER TO MEASURE THE AUREOLE. HERE ARE THE BASICS:

THE SUN SUBTENDS AN ANGLE OF ABOUT 0.5°.

THE SUN MOVES ITS DIAMETER IN 2 MINUTES.

POINT THE PHOTOMETER COLLIMATOR TUBE AT THE SUN, SECURE IT IN PLACE, AND ALLOW THE SUN TO DRIFT PAST THE COLLIMATOR'S FIELD OF VIEW.

COLLIMATOR TUBE HAS NO SHADOW WHEN IT IS POINTED DIRECTLY AT THE SUN.

HERE'S ONE WAY TO PLOT YOUR MEASUREMENTS:

TO COMPLETE THIS HALF OF SCAN, PLACE COLLIMATOR AHEAD OF WHERE SUN WILL DRIFT. (THIS TAKES PRACTICE.)

EARLY OR LATE PLOTS WILL BE WIDER THAN AT NOON.

SUNSHINE DURATION RECORDER

THE TOTAL TIME DURING A DAY WHEN THE SUN IS NOT BLOCKED BY CLOUDS IS AN IMPORTANT ENVIRONMENTAL PARAMETER IN AGRICULTURE AND STUDIES OF THE EFFECT OF CLOUDS ON THE EARTH'S TEMPERATURE. IN 1853 J.F. CAMPBELL INVENTED A SUNSHINE RECORDER:

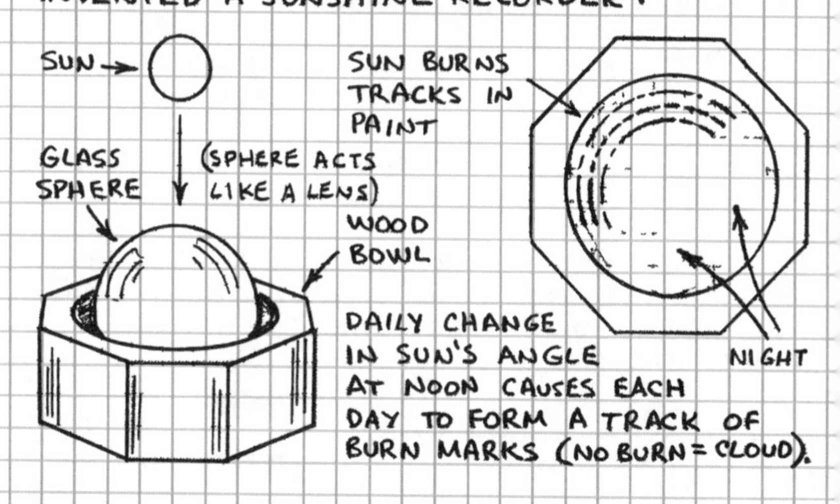

PAPER SUNSHINE RECORDER

SUNLIGHT DARKENS NEWSPRINT AND CAUSES SOME COLORED CONSTRUCTION PAPER TO FADE. PLACE STRIP OF BLUE OR RED CONSTRUCTION PAPER UNDER STRIP OF BLACK PAPER WITH SLOT CUT OUT TO PASS SUN LIGHT. MOVE SLOT SAME DISTANCE EACH MORNING. AFTER A WEEK, PAPER STRIP WILL HAVE SEVEN FADED RECTANGLES. THE MOST FADED RECTANGLES RECEIVED THE MOST SUNLIGHT.

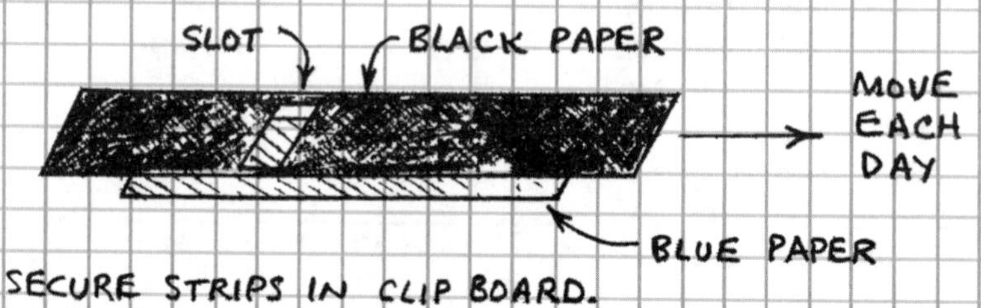

SECURE STRIPS IN CLIP BOARD.

POP BOTTLE SUNSHINE RECORDER

THIS SIMPLE APPARATUS INDICATES PASSAGE OF LARGE CLOUD MASSES AS UNFADED STRIPES ON BLUE CONSTRUCTION PAPER. RANDOM CLOUDS PASSING OVER SUN MAY CAUSE LESS FADING THAN A CLEAR DAY.

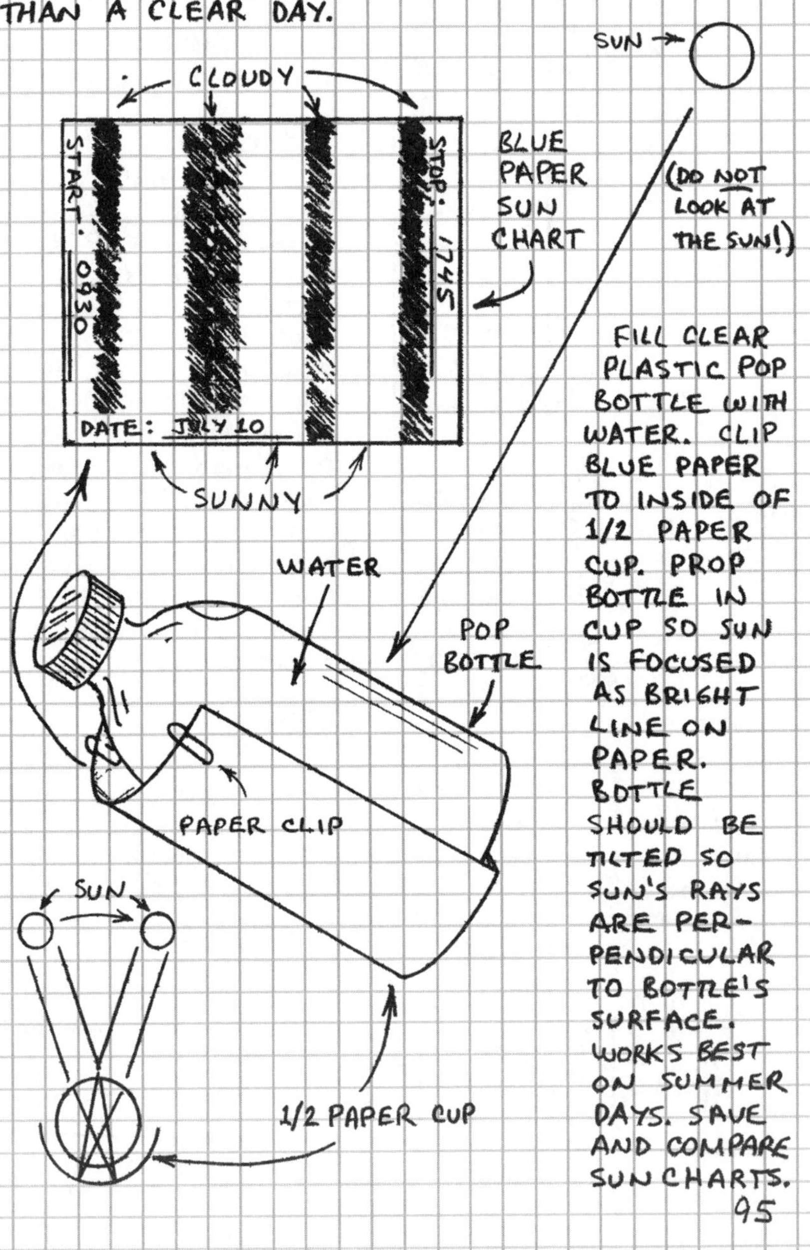

FILL CLEAR PLASTIC POP BOTTLE WITH WATER. CLIP BLUE PAPER TO INSIDE OF 1/2 PAPER CUP. PROP BOTTLE IN CUP SO SUN IS FOCUSED AS BRIGHT LINE ON PAPER. BOTTLE SHOULD BE TILTED SO SUN'S RAYS ARE PERPENDICULAR TO BOTTLE'S SURFACE. WORKS BEST ON SUMMER DAYS. SAVE AND COMPARE SUN CHARTS.

ELECTRONIC SUNSHINE RECORDER

THIS CIRCUIT MEASURES THE TOTAL TIME THE SUN SHINES DURING A DAY.

PC1-2 ARE CADMIUM SULFIDE PHOTORESISTORS.

PC1 IS CONCEALED FROM THE DIRECT SUN BY A SHADOW BAND. BOTH PC1 AND PC2 LOOK STRAIGHT UP. WITH PC1 IN SHADE AND PC2 IN SUN, ADJUST R1 UNTIL RELAY PULLS IN ("CLICK") AND CLOCK STARTS. SHADE PC2 AND RELAY SHOULD DROP OUT, STOPPING CLOCK. SET CLOCK TO 12:00:00 TO BEGIN. LOG TOTAL ELAPSED SUNSHINE TIME IN NOTEBOOK.

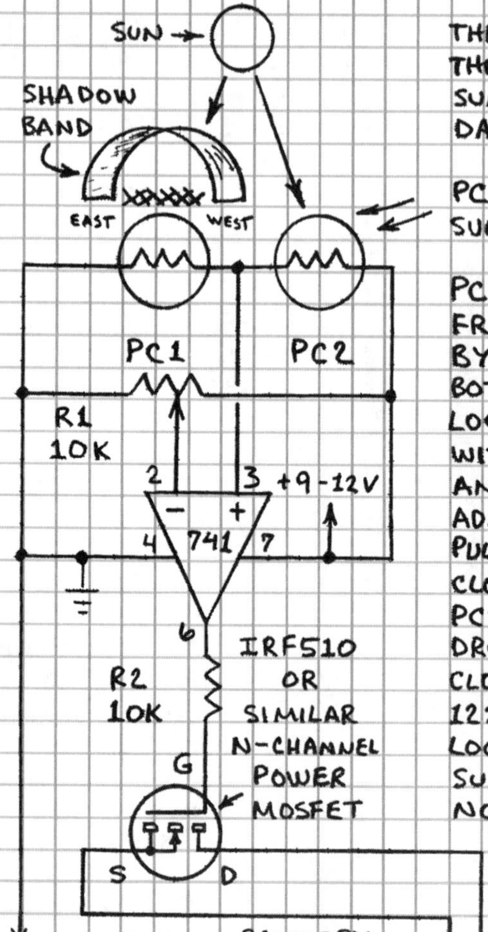

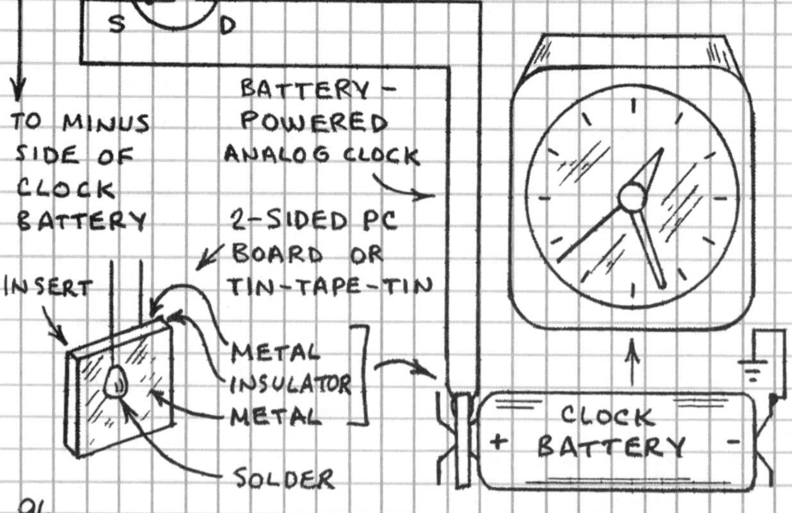

SUN
SHADOW BAND
EAST
WEST
PC1
PC2
R1 10K
2
3
+9-12V
4
741
7
6
R2 10K
IRF510 OR SIMILAR N-CHANNEL POWER MOSFET
G
S
D
TO MINUS SIDE OF CLOCK BATTERY
BATTERY-POWERED ANALOG CLOCK
2-SIDED PC BOARD OR TIN-TAPE-TIN
INSERT
METAL
INSULATOR
METAL
SOLDER
CLOCK BATTERY
+
-

III. COMMUNICATIONS PROJECTS

OVERVIEW

ELECTRONIC COMMUNICATIONS IS THE TRANSFER OF INFORMATION BY WIRE OR OPTICAL FIBER OR WIRELESS MEANS (RADIO, TELEVISION, MICROWAVE OR LIGHTWAVE).

THERE ARE MANY CATEGORIES OF ELECTRONIC COMMUNICATION. FOR INSTANCE, VOICE COMMUNICATIONS CAN BE 1-WAY AS IN A RADIO OR TELEVISION NEWS BROADCAST. OR VOICE COMMUNICATIONS CAN BE 2-WAY AS IN CONVERSATIONS VIA TELEPHONE, INTERCOM AND BOTH AMATEUR AND CITIZENS BAND RADIO. EXAMPLES OF NON-VOICE COMMUNICATION INCLUDE MORSE CODE, TELETYPEWRITER SIGNALS, COMPUTER DATA TRANSMISSION AND WILDLIFE TELEMETRY. RADIO CONTROL IS A FORM OF COMMUNICATION IN WHICH THE TRANSMITTED INFORMATION CONTROLS A REMOTE DEVICE SUCH AS A CAMERA, GARAGE DOOR OR MODEL BOAT OR PLANE.

CIRCUIT ASSEMBLY TIPS

THE CIRCUITS THAT FOLLOW CAN BE ASSEMBLED FROM READILY AVAILABLE SUPPLIES. YOU CAN USUALLY SUBSITUTE SIMILAR COMPONENTS IF THOSE SPECIFIED ARE UNAVAILABLE. FOR INSTANCE, A 25,000 (50K) OHM POTENTIOMETER CAN BE SUBSTITUTED FOR A 10,000 (10K) UNIT. BE SURE TO BYPASS THE POWER SUPPLY PINS OF OPERATIONAL AND POWER AMPLIER ICs (TIE THEM TO GROUND WITH A 0.1μF CAPACITOR CONNECTED CLOSE TO THE IC). THIS WILL HELP PREVENT UNWANTED OSCILLATION. FOR ADDITIONAL INFORMATION SEE "GETTING STARTED IN ELECTRONICS" (RADIO SHACK, 1983) AND OTHER BOOKS IN THIS SERIES.

CONNECTED COMMUNICATION LINKS

CONNECTED COMMUNICATION LINKS ARE THOSE IN WHICH TWO OR MORE STATIONS ARE LINKED BY A WIRE, CABLE OR WAVEGUIDE.

ADVANTAGES INCLUDE RELIABILITY, LOW NOISE AND SIMPLE ELECTRONICS. HOWEVER, CONNECTED LINKS REQUIRE RIGHT-OF-WAY AND CAN BE VERY EXPENSIVE TO INSTALL. FURTHERMORE, ONLY CONNECTED STATIONS CAN COMMUNICATE.

SINGLE WIRE

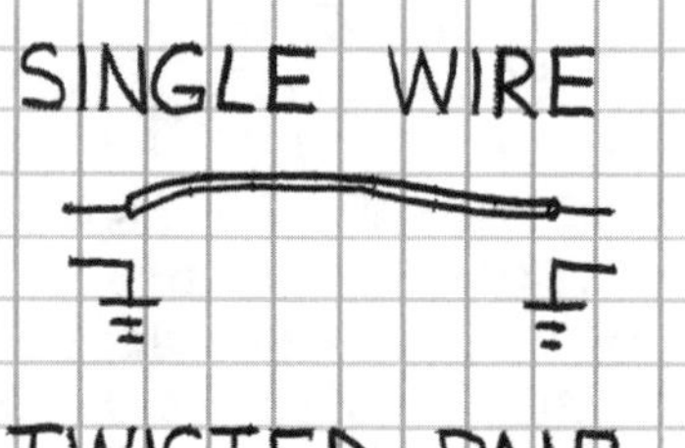

TELEGRAPH LINKS. GROUND REQUIRED AT EACH END.

TWISTED PAIR

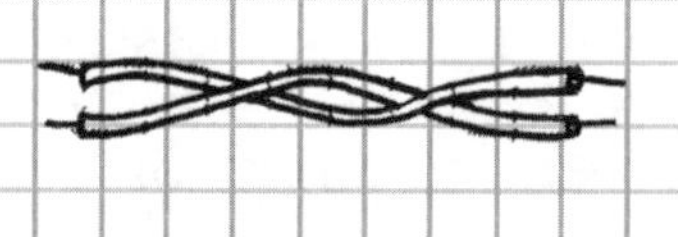

TELEPHONES (UP TO 15 CHANNELS) AND DIGITAL DATA TRANSMISSION.

COAXIAL CABLE

CAN CARRY UP TO 90,000 VOICE CHANNELS.

HOLLOW WAVEGUIDE

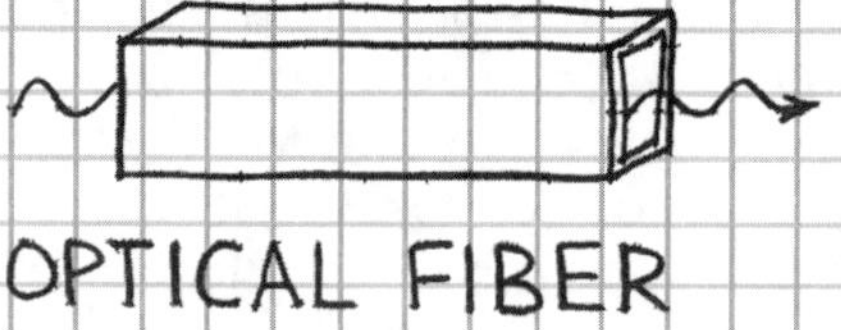

CAN CARRY MICROWAVE SIGNAL MODULATED WITH 100,000 + VOICE CHANNELS.

OPTICAL FIBER

CAN CARRY LIGHTWAVE MODULATED WITH 100,000 OR MORE VOICE CHANNELS.

WIRELESS COMMUNICATION LINKS

WIRELESS COMMUNICATIONS LINKS ARE THOSE IN WHICH INFORMATION IS SENT TO ONE OR MORE RECEIVERS BY MEANS OF A MODULATED ELECTRO-MAGNETIC WAVE.

ADVANTAGES INCLUDE LONG DISTANCE COMMUNICATION, TRANSMISSION TO AND FROM LAND, AIR AND SPACE VEHICLES AND BOTH DIRECTIONAL AND NON-DIRECTIONAL TRANSMISSION. SUBJECT TO INTERFERING NOISE.

RADIO

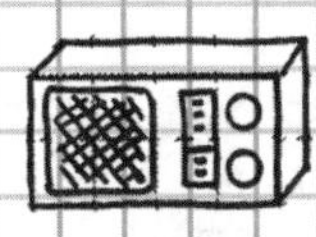

BROADCAST AND SHORTWAVE RADIO. ALSO AMATEUR RADIO, CITIZENS BAND, MOBILE, ETC.

VHF

TELEVISION AND FM RADIO. ALSO AIRCRAFT, AMATEUR RADIO, MOBILE, SPACE, ETC.

UHF

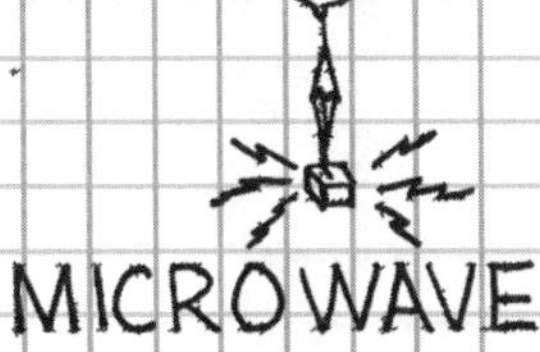

WEATHER BALLOONS, TELEVISION, MOBILE, NAVIGATION, AMATEUR, SATELLITE, DEEP SPACE, ETC.

MICROWAVE

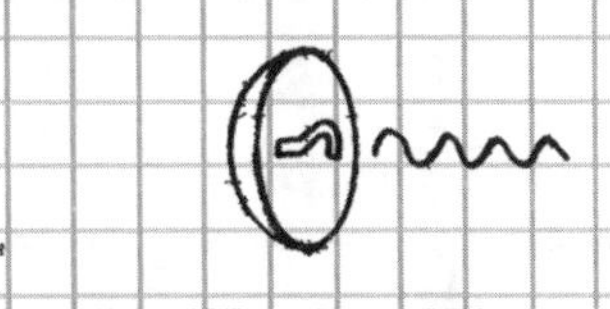

COMMUNICATIONS SATELLITE, LONG DISTANCE TELEPHONE, NAVIGATION, AMATEUR, ETC.

LIGHTWAVE

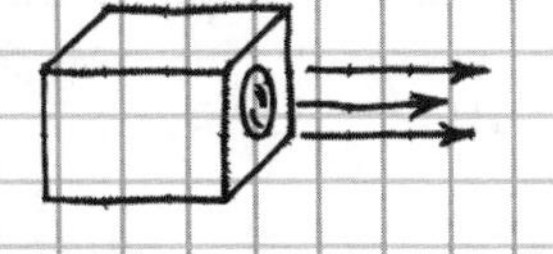

LINE-OF-SIGHT COMPUTER DATA TRANSMISSION AND VOICE LINKS.

ELECTROMAGNETIC RADIATION

ELECTROMAGNETIC RADIATION IS ENERGY IN THE FORM OF A WAVE OF OSCILLATING ELECTRIC AND MAGNETIC FIELDS. THE WAVE TRAVELS THROUGH A VACUUM AT A VELOCITY OF 2.998×10^8 METERS PER SECOND (186,284 MILES PER SECOND). THE WAVELENGTH OF AN ELECTROMAGNETIC WAVE DETERMINES ITS PROPERTIES. X-RAYS, INFRARED, MICROWAVES, RADIO WAVES AND LIGHT ARE ELECTROMAGNETIC RADIATION.

ELECTROMAGNETIC SPECTRUM

nm = NANOMETER (1nm = .000 000 001 METER)
μ = MICROMETER (1 μ = .000 001 METER)
mm = MILLIMETER (1mm = .001 METER)
m = METER (1 m = 39.37 INCHES)
km = KILOMETER (1 km = 1,000 METERS)

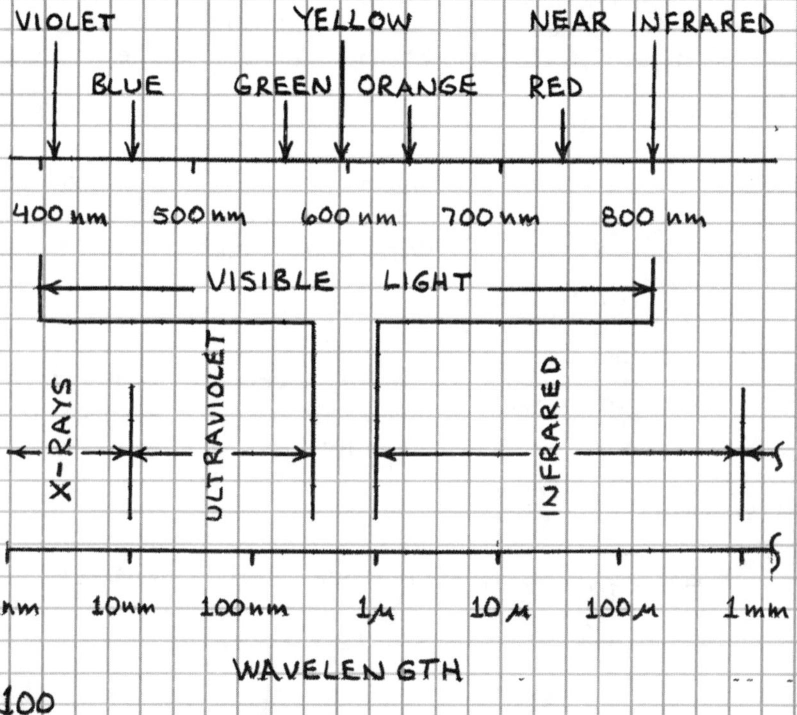

VIOLET
BLUE
GREEN
YELLOW
ORANGE
RED
NEAR INFRARED

400 nm 500 nm 600 nm 700 nm 800 nm

VISIBLE LIGHT

X-RAYS
ULTRAVIOLET
INFRARED

nm 10nm 100nm 1μ 10 μ 100 μ 1 mm

WAVELENGTH

WAVELENGTH VS FREQUENCY

THE FREQUENCY OF AN ELECTROMAGNETIC WAVE IS THE NUMBER OF CYCLES THAT OCCUR IN ONE SECOND.

1 CYCLE

1 CYCLE/SECOND = 1 HERTZ (Hz)

TIME

WAVELENGTH

IF EITHER THE FREQUENCY OR LENGTH OF A WAVE IS KNOWN, THE UNKNOWN VALUE CAN BE CALCULATED:

FREQUENCY (Hz) = C / WAVELENGTH (λ)

WAVELENGTH (λ) = C / FREQUENCY (Hz)

$C = 3 \times 10^8$ METERS PER SECOND

RADIO WAVES

MICROWAVES

ULTRA-HIGH FREQUENCY (UHF)

VERY-HIGH FREQUENCY (VHF)

HIGH-FREQUENCY (HF)

MEDIUM-FREQUENCY (MF)

10 mm 100 mm 1 m 10 m 100 m 1 Km

WAVELENGTH

INTERNATIONAL MORSE CODE

IN 1836, SAMUEL F. B. MORSE BUILT THE FIRST WORKING TELEGRAPH. HE ALSO DEVISED A CODE THAT PERMITTED TELEGRAPH OPERATORS TO EXCHANGE INFORMATION. HIS CODE IS STILL USED BY TELEGRAPH, RADIO AND SIGNAL LIGHT OPERATORS. HERE IT IS:

A	· —	N	— ·	1	· — — — —
B	— · · ·	O	— — —	2	· · — — —
C	— · — ·	P	· — — ·	3	· · · — —
D	— · ·	Q	— — · —	4	· · · · —
E	·	R	· — ·	5	· · · · ·
F	· · — ·	S	· · ·	6	— · · · ·
G	— — ·	T	—	7	— — · · ·
H	· · · ·	U	· · —	8	— — — · ·
I	· ·	V	· · · —	9	— — — — ·
J	· — — —	W	· — —	0	— — — — —
K	— · —	X	— · · —	.	· — · — · —
L	· — · ·	Y	— · — —	?	· · — — · ·
M	— —	Z	— — · ·	-	— · · · · —

THE CODE INCLUDES MANY ADDITIONAL PUNCTUATION MARKS, PHRASES AND ABBREVIATIONS.

LEARNING THE CODE

THINK OF THE CODE AS SOUNDS, NOT DOTS AND DASHES. SAY "DIT" FOR DOT AND "DAH" FOR DASH. THUS A IS "DIT DAH" OR SIMPLY "DIDAH." B IS "DAHDIDIDIT." C IS "DAHDIDAHDIT." A CODE PRACTICE OSCILLATOR CAN HELP YOU LEARN THE CODE. EVEN BETTER IS THE CASSETTE TAPE INCLUDED WITH THE "TUNE IN THE WORLD WITH HAM RADIO" KIT AVAILABLE FROM THE AMERICAN RADIO RELAY LEAGUE (ARRL) IN NEWINGTON, CT 06111. THE TEXT SUPPLIED WITH THE KIT IS AN EXCELLENT INTRODUCTION TO THE WORLD OF AMATEUR RADIO. IT COVERS ELECTRICAL THEORY, EQUIPMENT, ANTENNAS, ETC.

CODE PRACTICE OSCILLATORS

A RADIO TRANSMITTER REQUIRES LESS POWER TO TRANSMIT CODE THAN VOICE. MOREOVER, CODE CAN BE UNDERSTOOD WHEN THE SIGNAL IS VERY FAINT OR WHEN STATIC IS SO SEVERE THAT VOICE IS UNINTELLIGIBLE. THESE CPO'S WILL HELP YOU LEARN CODE.

PIEZOBUZZER CPO

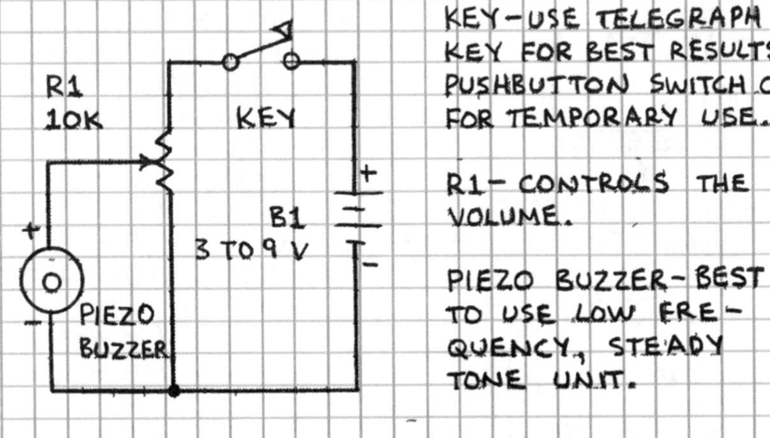

KEY–USE TELEGRAPH KEY FOR BEST RESULTS. PUSHBUTTON SWITCH OK FOR TEMPORARY USE.

R1–CONTROLS THE VOLUME.

PIEZO BUZZER–BEST TO USE LOW FREQUENCY, STEADY TONE UNIT.

INTEGRATED CIRCUIT CPO

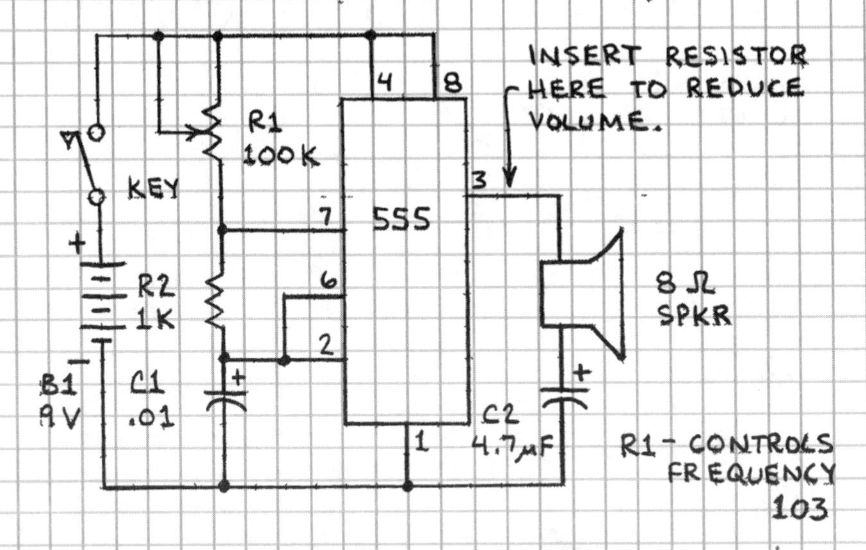

INSERT RESISTOR HERE TO REDUCE VOLUME.

R1–CONTROLS FREQUENCY

ELECTROMAGNETIC TELEGRAPH

THERE ARE MANY WAYS TO MAKE SIMPLE TELEGRAPHS. FOR EXAMPLE, THE CODE PRACTICE OSCILLATORS ON THE PREVIOUS PAGE CAN BE USED IN A SOLID-STATE TELEGRAPH SYSTEM. THE COMPONENTS OF A DO-IT-YOURSELF ELECTROMAGNETIC TELEGRAPH ARE GIVEN HERE. YOU CAN BUILD THE TELEGRAPH ON THE FACING PAGE IN A FEW HOURS.

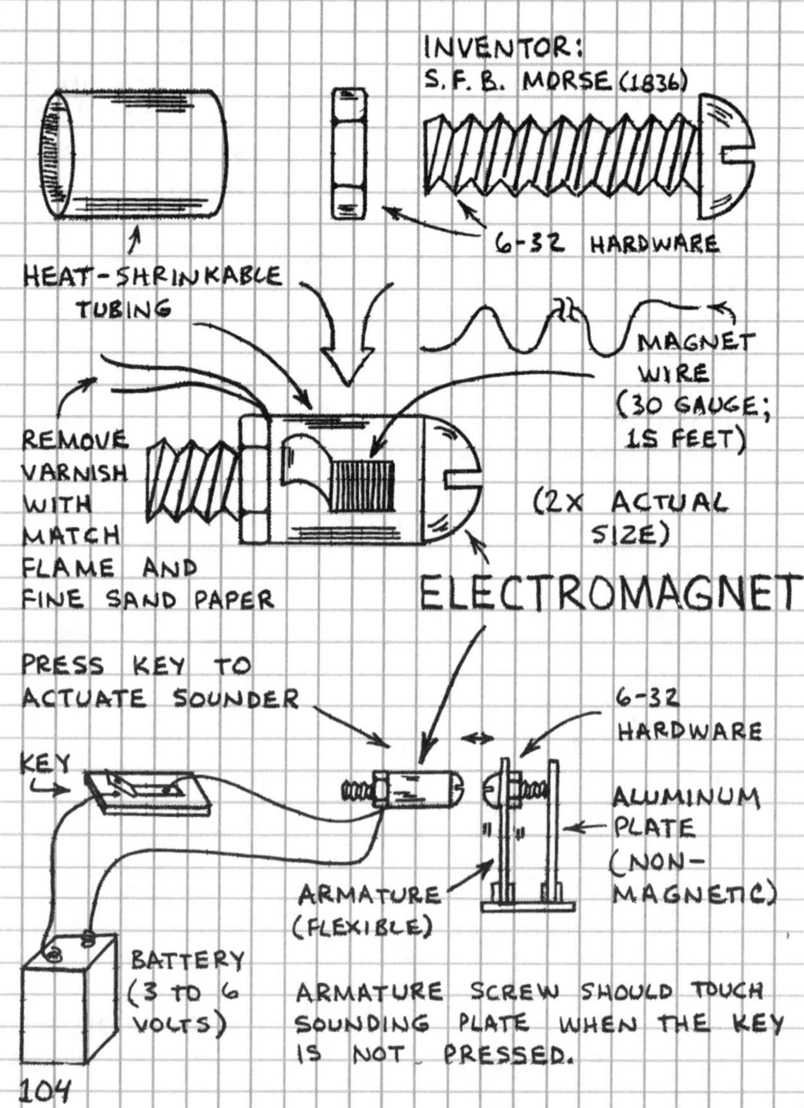

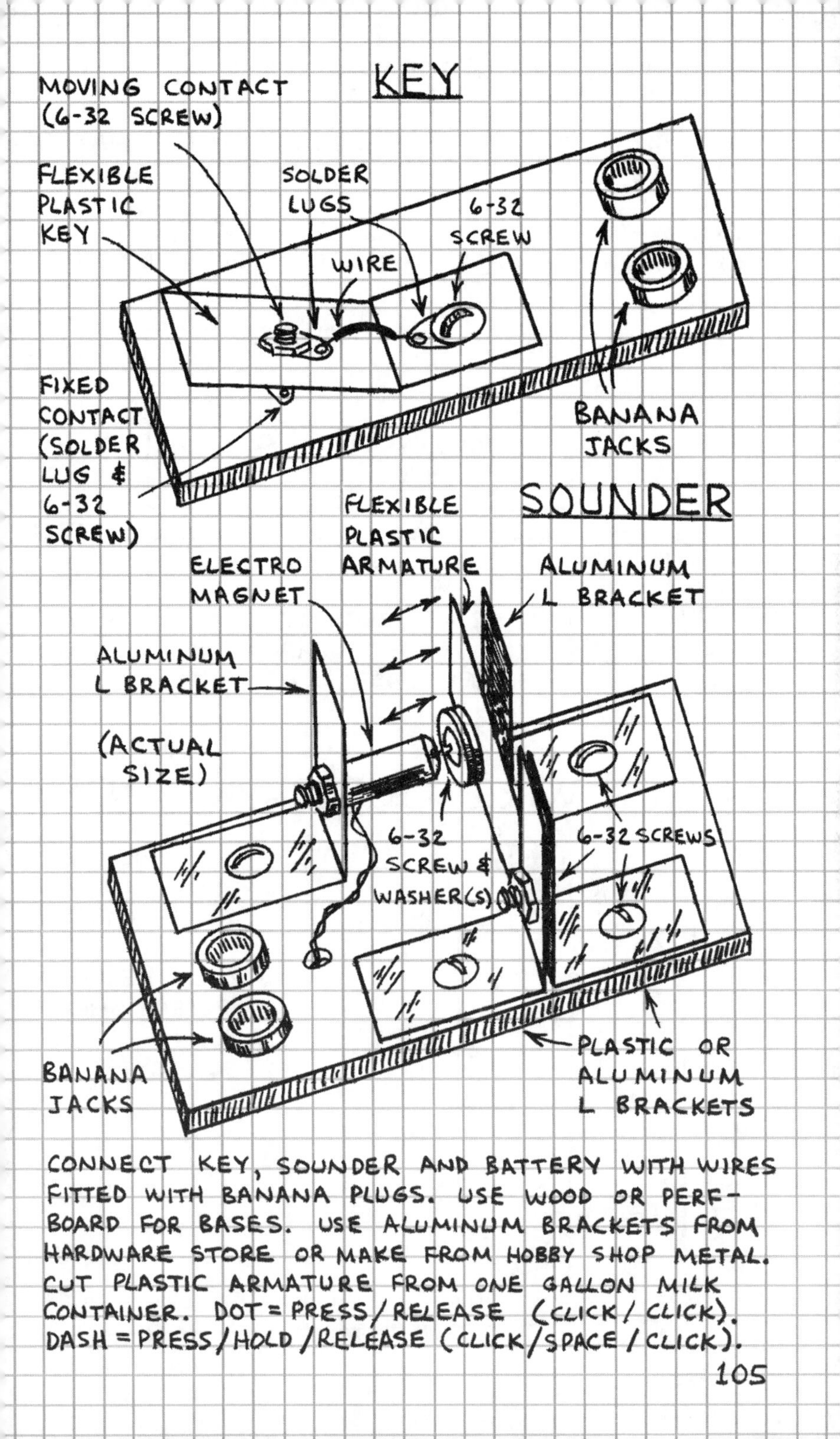

CONNECT KEY, SOUNDER AND BATTERY WITH WIRES FITTED WITH BANANA PLUGS. USE WOOD OR PERF-BOARD FOR BASES. USE ALUMINUM BRACKETS FROM HARDWARE STORE OR MAKE FROM HOBBY SHOP METAL. CUT PLASTIC ARMATURE FROM ONE GALLON MILK CONTAINER. DOT = PRESS/RELEASE (CLICK/CLICK). DASH = PRESS/HOLD/RELEASE (CLICK/SPACE/CLICK).

SOLID-STATE TELEGRAPHS

TRANSISTORS AND INTEGRATED CIRCUITS MAKE POSSIBLE VERY SENSITIVE TELEGRAPH SYSTEMS.

CAUTION. NEVER INSTALL TELEGRAPH, INTERCOM OR TELEPHONE WIRES NEAR OUTDOOR POWER LINES.

SIMPLE SOLID-STATE TELEGRAPH

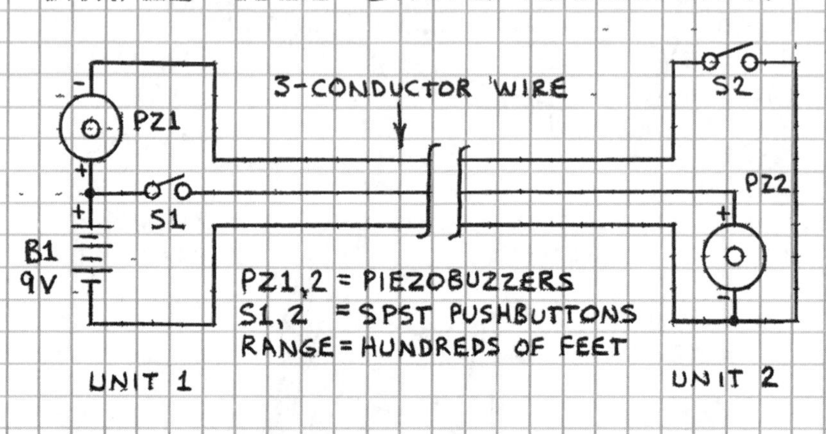

1-OR 2-WIRE TELEGRAPH SENDER

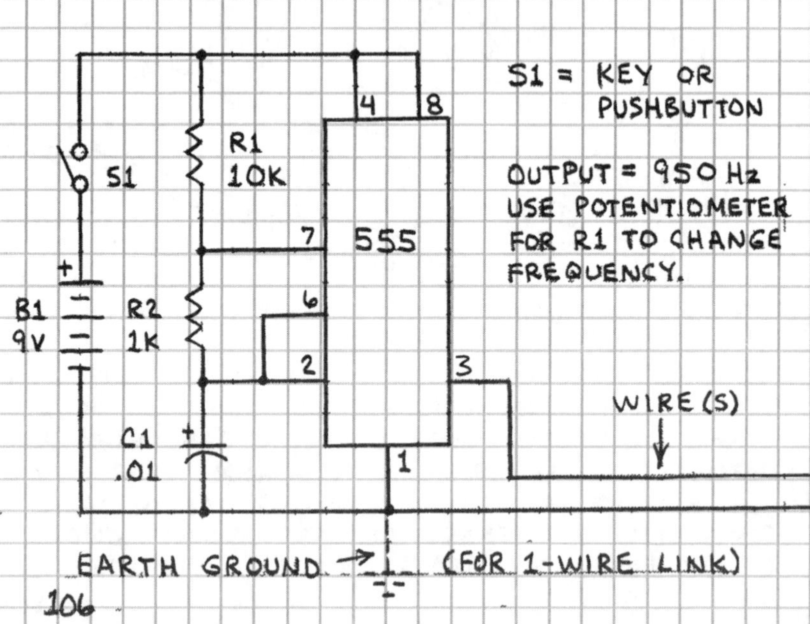

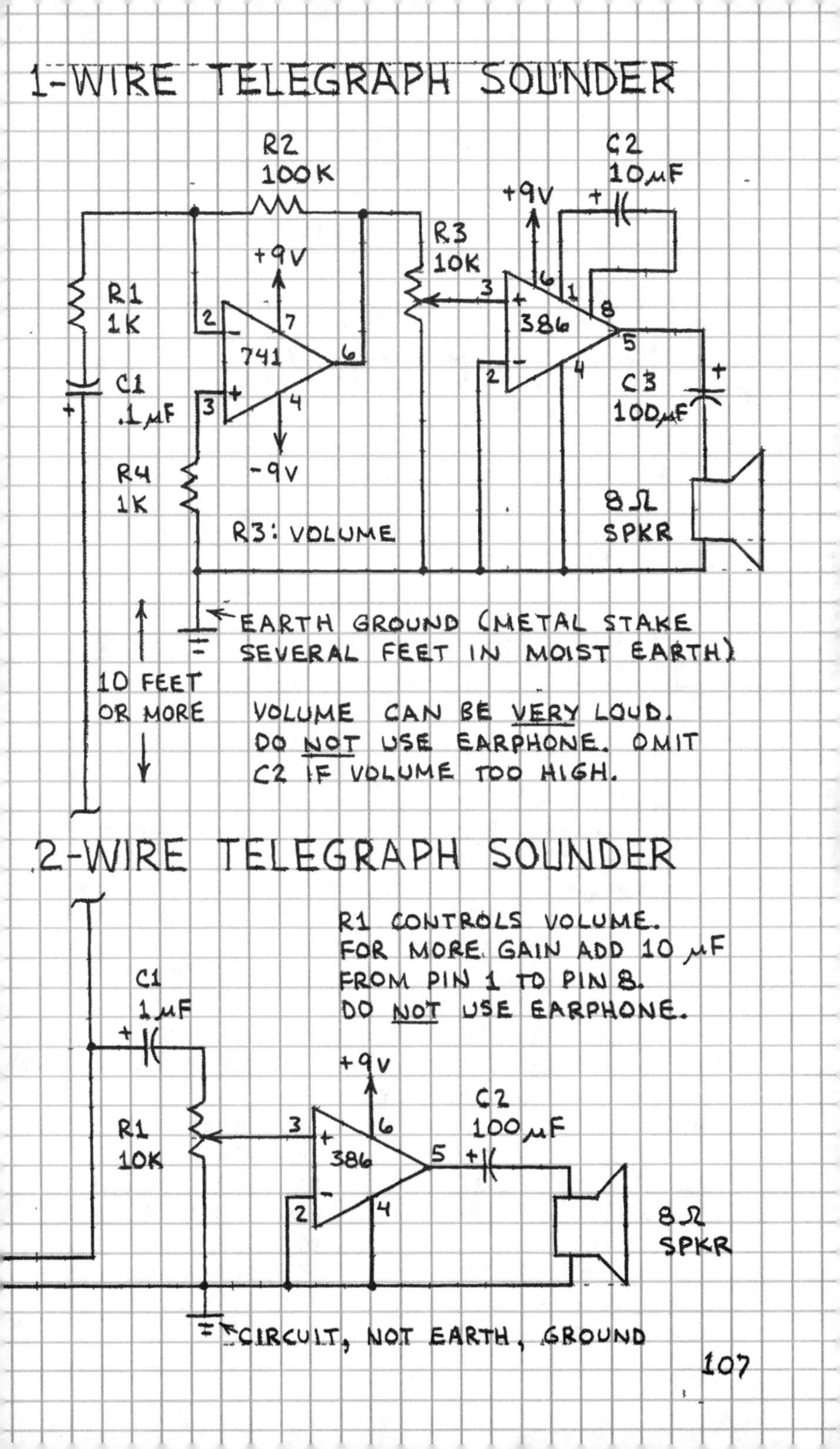
1-WIRE TELEGRAPH SOUNDER
R2
100K
R1
1K
C1
.1µF
+
+9V
2
7
741
6
3
+
4
-9V
R4
1K
R3
10K
R3: VOLUME
+9V
6
1
3
+
386
8
5
2
-
4
C2
10µF
+
C3
100µF
+
8Ω
SPKR
EARTH GROUND (METAL STAKE SEVERAL FEET IN MOIST EARTH)
10 FEET OR MORE
VOLUME CAN BE VERY LOUD. DO NOT USE EARPHONE. OMIT C2 IF VOLUME TOO HIGH.
2-WIRE TELEGRAPH SOUNDER
R1 CONTROLS VOLUME.
FOR MORE GAIN ADD 10 µF FROM PIN 1 TO PIN 8.
DO NOT USE EARPHONE.
C1
1µF
+
R1
10K
+9V
3
+
6
386
5
2
-
4
C2
100µF
+
8Ω
SPKR
CIRCUIT, NOT EARTH, GROUND

TELEPHONE RECEIVER

A SIMPLE TELEPHONE RECEIVER IS EASILY MADE FROM READILY AVAILABLE MATERIALS:

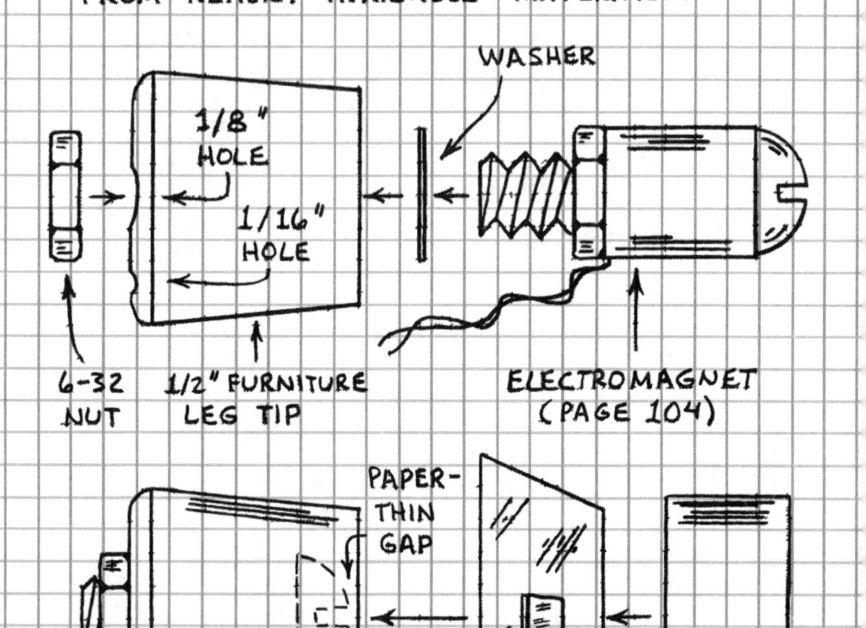

PAPER-THIN GAP

STEEL ARMATURE

PLASTIC FILM

5/8" BRASS TUBING (HOBBY SHOP)

ARMATURE IS 3/16" SQUARE, 1/32" THICK STEEL (SCRAP OR CUT FROM SHEET). ATTACH TO PLASTIC WITH DOUBLE-SIDED TAPE.

INVENTOR.
PROF. A. G. BELL (1876)

ACTUAL SIZE

10 Ω

ADD 10 OHM RESISTOR. CONNECT LEADS TO BATTERY-POWERED RADIO PHONE JACK TO TEST. VOLUME WILL BE LOW SINCE COIL RESISTANCE IS ONLY 1.56 OHMS.

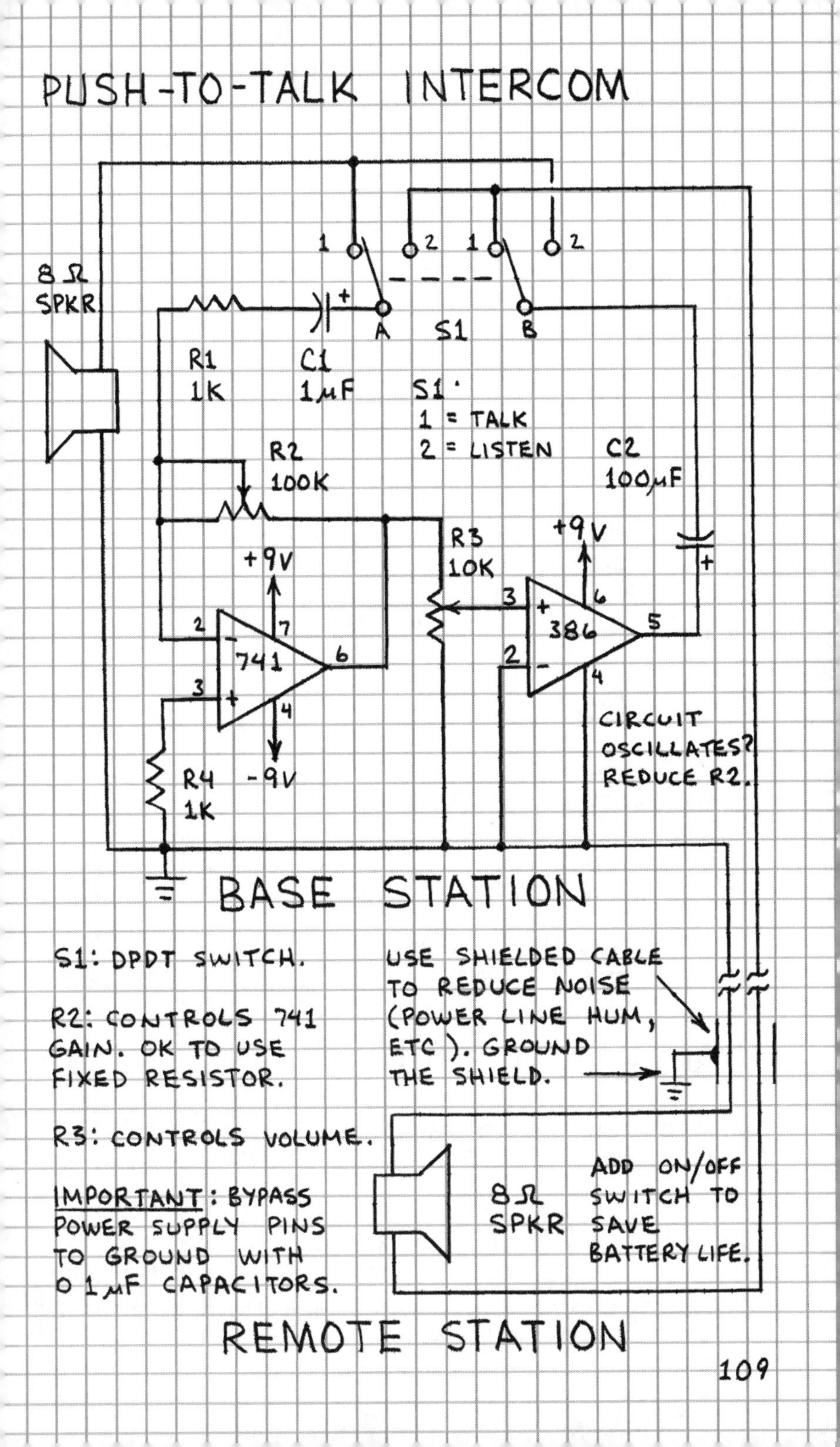
PUSH-TO-TALK INTERCOM
8 Ω
SPKR
R1
1K
C1
1µF
A
S1
B
1
2
1
2
S1:
1 = TALK
2 = LISTEN
R2
100K
C2
100µF
+9V
R3
10K
+9V
741
386
R4
1K
-9V
CIRCUIT OSCILLATES? REDUCE R2.
BASE STATION
S1: DPDT SWITCH.
R2: CONTROLS 741 GAIN. OK TO USE FIXED RESISTOR.
R3: CONTROLS VOLUME.
IMPORTANT: BYPASS POWER SUPPLY PINS TO GROUND WITH 0.1µF CAPACITORS.
USE SHIELDED CABLE TO REDUCE NOISE (POWER LINE HUM, ETC). GROUND THE SHIELD.
8 Ω
SPKR
ADD ON/OFF SWITCH TO SAVE BATTERY LIFE.
REMOTE STATION

LIGHTWAVE COMMUNICATIONS

1880 - ALEXANDER GRAHAM BELL INVENTED THE PHOTOPHONE, A DEVICE FOR SENDING VOICE OVER A BEAM OF SUNLIGHT.

1880 - BELL AND SUMNER TAINTER SENT VOICE MESSAGES OVER A 213 METER PATH.

1966 - K C. KAO PROPOSED LONG DISTANCE OPTICAL FIBER COMMUNICATIONS.

MODULATION

A LIGHTWAVE CAN CARRY DIGITAL DATA OR ANALOG INFORMATION SUCH AS VOICE. SHOWN BELOW ARE SOME WAYS IN WHICH A LIGHT WAVE CAN BE ANALOG MODULATED.

ANALOG SIGNAL

TYPICAL ANALOG SIGNAL (TEMPERATURE, TONE, ETC.).

AMPLITUDE

ANALOG SIGNAL CONTROLS INTENSITY OF LIGHT.

PULSE AMPLITUDE

ANALOG SIGNAL CONTROLS INTENSITY OF PULSES.

PULSE FREQUENCY

ANALOG SIGNAL CONTROLS FREQUENCY OF PULSES.

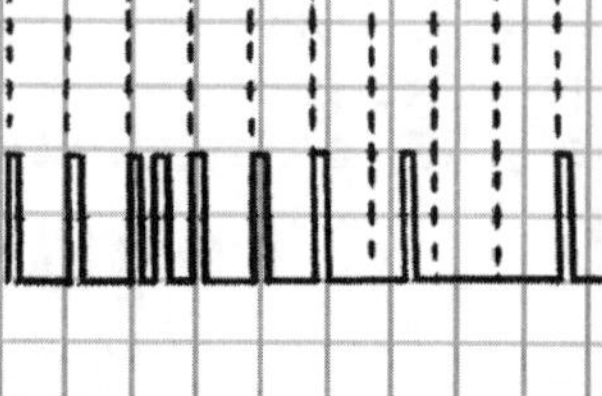

LIGHT SOURCES

MANY LIGHT SOURCES CAN BE USED IN LIGHTWAVE COMMUNICATION SYSTEMS. AMONG THE EASIEST TO USE ARE:

1. SUNLIGHT—
USED IN THE FIRST LIGHTWAVE COMMUNICATORS AND STILL VERY EASY TO USE.

2. INCANDESCENT LAMP—
LAMPS WITH SMALL FILAMENTS CAN BE VOICE MODULATED. NOT SUITABLE FOR HIGH FREQUENCY SIGNALS.

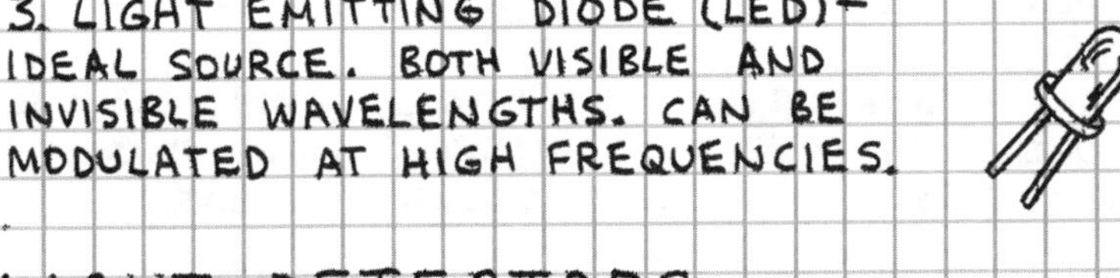

3. LIGHT EMITTING DIODE (LED)—
IDEAL SOURCE. BOTH VISIBLE AND INVISIBLE WAVELENGTHS. CAN BE MODULATED AT HIGH FREQUENCIES.

LIGHT DETECTORS

DETECTORS FOR LIGHTWAVE COMMUNICATION LINKS ARE USUALLY SOLID-STATE DEVICES. AMONG THE MOST COMMONLY USED ARE:

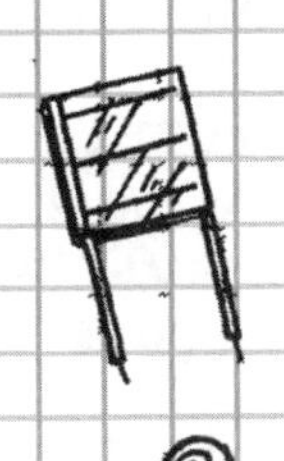

1. SOLAR CELL—
INEXPENSIVE AND EASY TO USE. PEAK SENSITIVITY IS ~880 nm. CAN BE USED FROM ~450 nm TO 1100 nm.

2. PHOTOTRANSISTOR—
FASTER AND MORE SENSITIVE THAN SOLAR CELLS. SAME SPECTRAL RESPONSE. EXTERNAL LENS HELPFUL.

3. LIGHT EMITTING DIODE—
AN LED CAN DETECT THE EMISSION FROM A SIMILAR LED. RED AND NEAR-INFRARED LEDS WORK BEST AS DETECTORS.

LIGHTWAVE SYSTEMS

MODULATED LIGHTWAVES CAN BE SENT THROUGH AIR (FREE SPACE) OR ULTRA-CLEAR OPTICAL FIBERS.

LINK	ADVANTAGES	DISADVANTAGES
FREE SPACE	1. NO LICENSE 2 PRIVACY 3 JAM PROOF	1. HARD TO ALIGN 2 LINE OF SIGHT 3. RAIN AND FOG
FIBER	1. VERY LOW NOISE 2. LIGHTNING PROOF 3 SECURITY	1. INSTALLATION 2. HIGHER COST 3. HARD TO SPLICE

FREE SPACE LINKS

SHORT RANGE SYSTEMS (0 TO 10 FEET) VERY EASY TO DESIGN AND ALIGN. LONGER RANGES USUALLY REQUIRE EXTERNAL LENSES AND TRIPODS.

ALIGNMENT METHODS INCLUDE.

1. REFLECTOR — USE RED LED AND PLACE BIKE REFLECTOR NEXT TO RECEIVER. POINT TRANSMITTER AT REFLECTOR.

2. TELESCOPE — BORESIGHT A SMALL TELESCOPE MOUNTED ON THE TRANSMITTER.

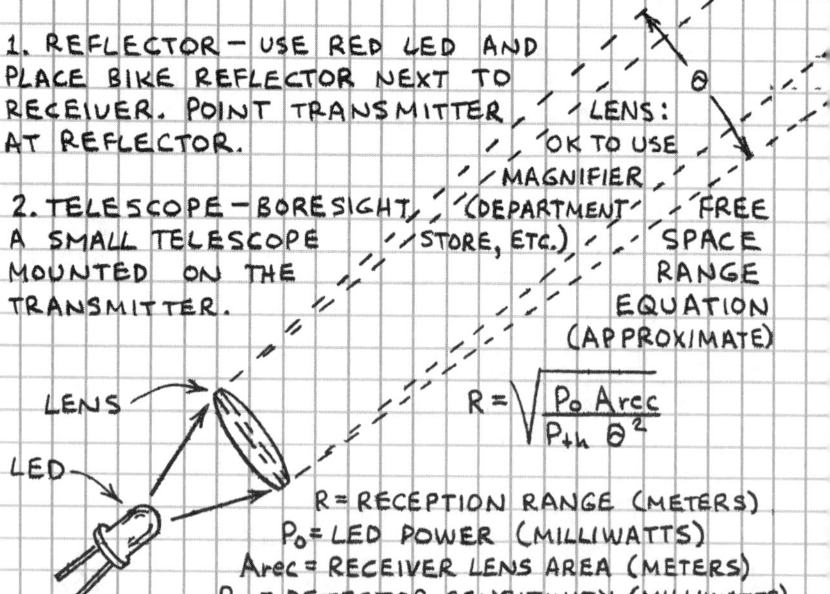

LENS: OK TO USE MAGNIFIER (DEPARTMENT STORE, ETC.)

θ

LENS

LED

FREE SPACE RANGE EQUATION (APPROXIMATE)

$$R = \sqrt{\frac{P_o A_{rec}}{P_{th}\,\theta^2}}$$

R = RECEPTION RANGE (METERS)
P_o = LED POWER (MILLIWATTS)
A_{rec} = RECEIVER LENS AREA (METERS)
P_{th} = DETECTOR SENSITIVITY (MILLIWATTS)
θ = LED BEAM DIVERGENCE (RADIANS)

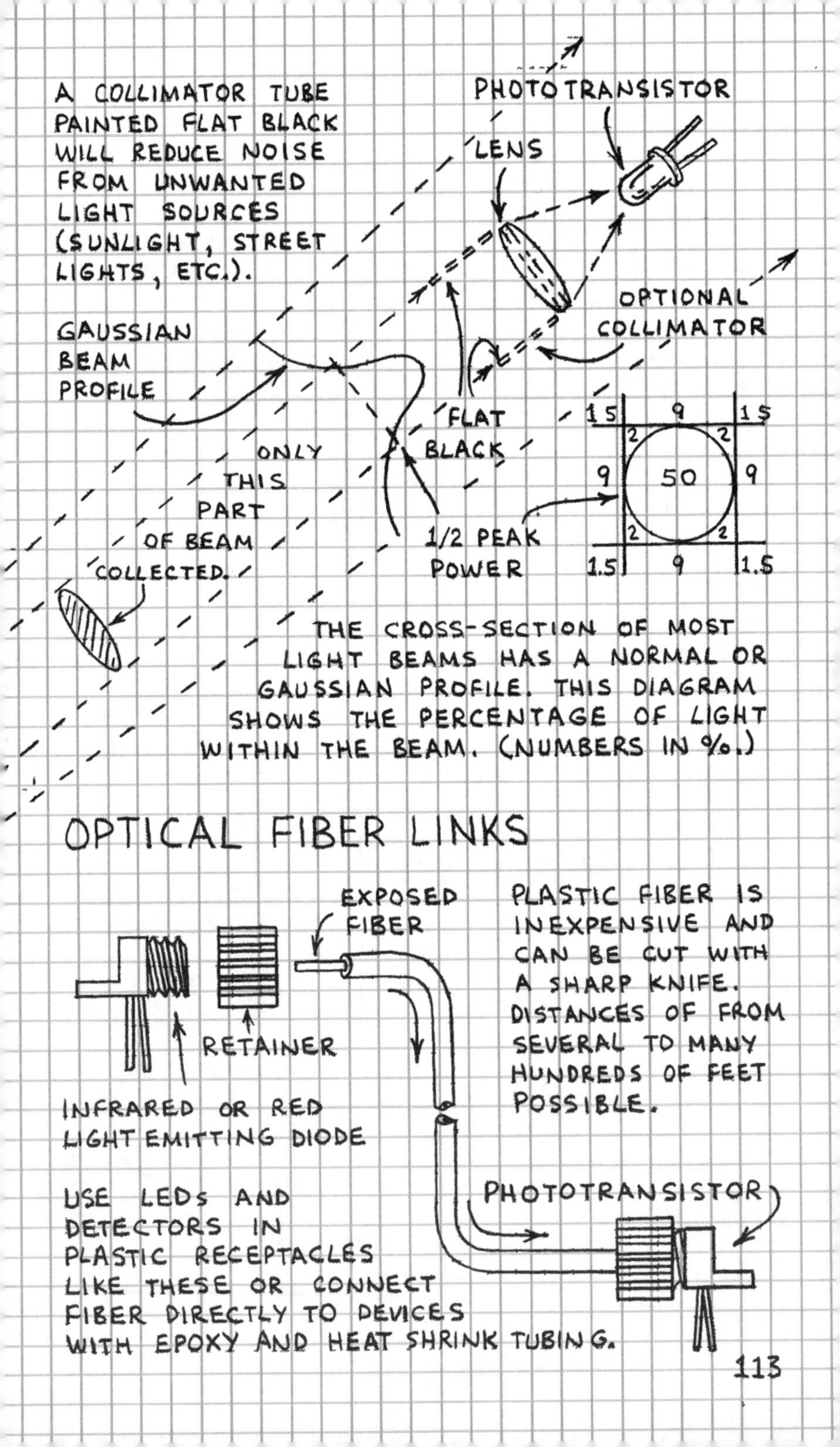

A COLLIMATOR TUBE PAINTED FLAT BLACK WILL REDUCE NOISE FROM UNWANTED LIGHT SOURCES (SUNLIGHT, STREET LIGHTS, ETC.).

THE CROSS-SECTION OF MOST LIGHT BEAMS HAS A NORMAL OR GAUSSIAN PROFILE. THIS DIAGRAM SHOWS THE PERCENTAGE OF LIGHT WITHIN THE BEAM. (NUMBERS IN %.)

OPTICAL FIBER LINKS

PLASTIC FIBER IS INEXPENSIVE AND CAN BE CUT WITH A SHARP KNIFE. DISTANCES OF FROM SEVERAL TO MANY HUNDREDS OF FEET POSSIBLE.

USE LEDs AND DETECTORS IN PLASTIC RECEPTACLES LIKE THESE OR CONNECT FIBER DIRECTLY TO DEVICES WITH EPOXY AND HEAT SHRINK TUBING.

ELECTRONIC PHOTOPHONE

AFTER HE INVENTED THE PHOTOPHONE IN 1880, ALEXANDER GRAHAM BELL INVENTED THE ELECTRIC PHOTOPHONE. IN THE NON-ELECTRIC PHOTOPHONE A BEAM OF SUNLIGHT WAS DIRECTLY MODULATED BY VOICE PRESSURE AGAINST A FLEXIBLE MIRROR OR MOVABLE GRATING. IN THE ELECTRIC PHOTOPHONE SUNLIGHT WAS MODULATED BY A MIRROR ATTACHED TO A TELEPHONE RECEIVER. SHOWN HERE IS A MODERN VERSION OF THE ELECTRIC PHOTOPHONE.

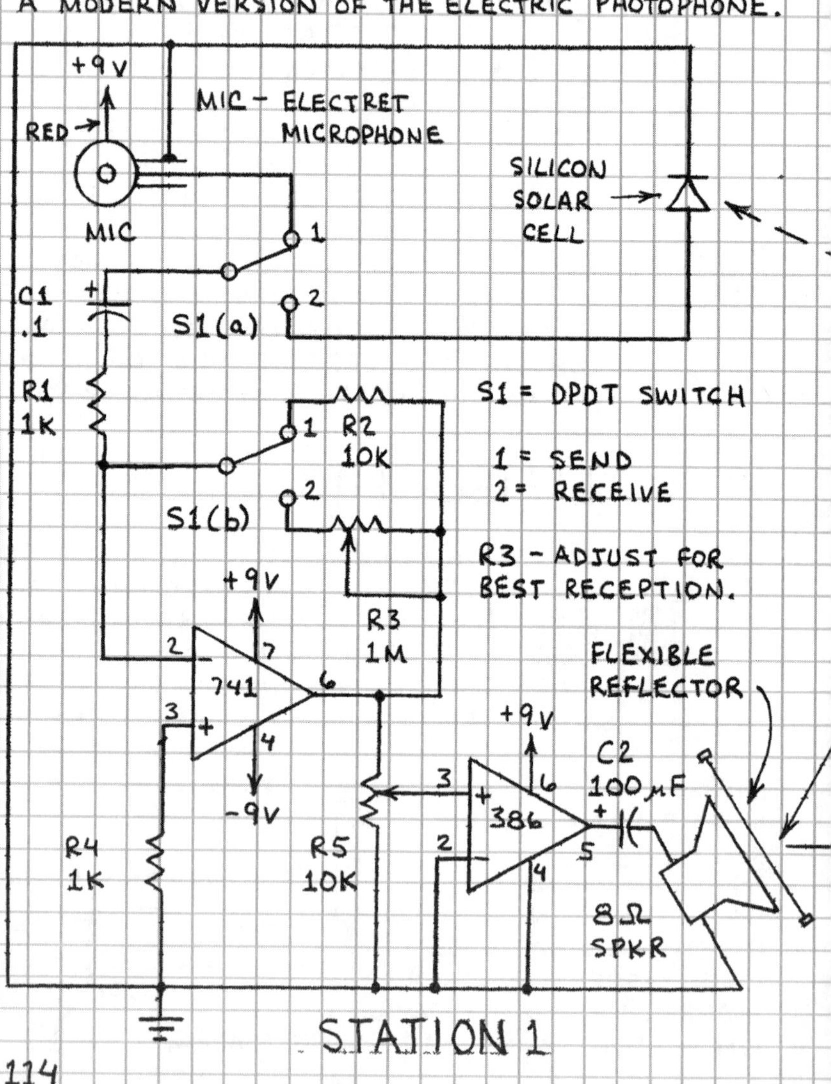

KEEP BATTERY LEADS SHORT AND CONNECT 0.1 μF CAPACITORS FROM POWER SUPPLY PINS OF EACH CHIP TO GROUND.

SUN

IMPORTANT; THE SPEAKERS MAY EMIT VERY LOUD SOUNDS. DO NOT PLACE YOUR EARS CLOSE TO EITHER SPEAKER.

FLEXIBLE REFLECTOR IS ALUMINIZED MYLAR OR HEAVY DUTY ALUMINUM FOIL STRETCHED OVER SPEAKER OR HOLE IN BOX IN WHICH SPEAKER IS INSTALLED. USE ALUMINIZED MYLAR FROM EMERGENCY BLANKET OR PACKAGING MATERIAL.

CAUTION. BOTH OPERATORS MUST WEAR SUNGLASSES AND AVOID STARING AT REFLECTED SUNLIGHT!

FLEXIBLE REFLECTOR

USE TRIPODS FOR BEST RESULTS. REFLECTED SUNLIGHT FROM FLEXIBLE REFLECTOR SHOULD FORM A DISTINCT SPOT WHEN DIRECTED AGAINST A NEARBY WALL.

SOLAR CELL

(SEE FACING PAGE)

NOTE THAT THE SPEAKERS FUNCTION AS SOUND SOURCE IN RECEIVE MODE.

STATION 2

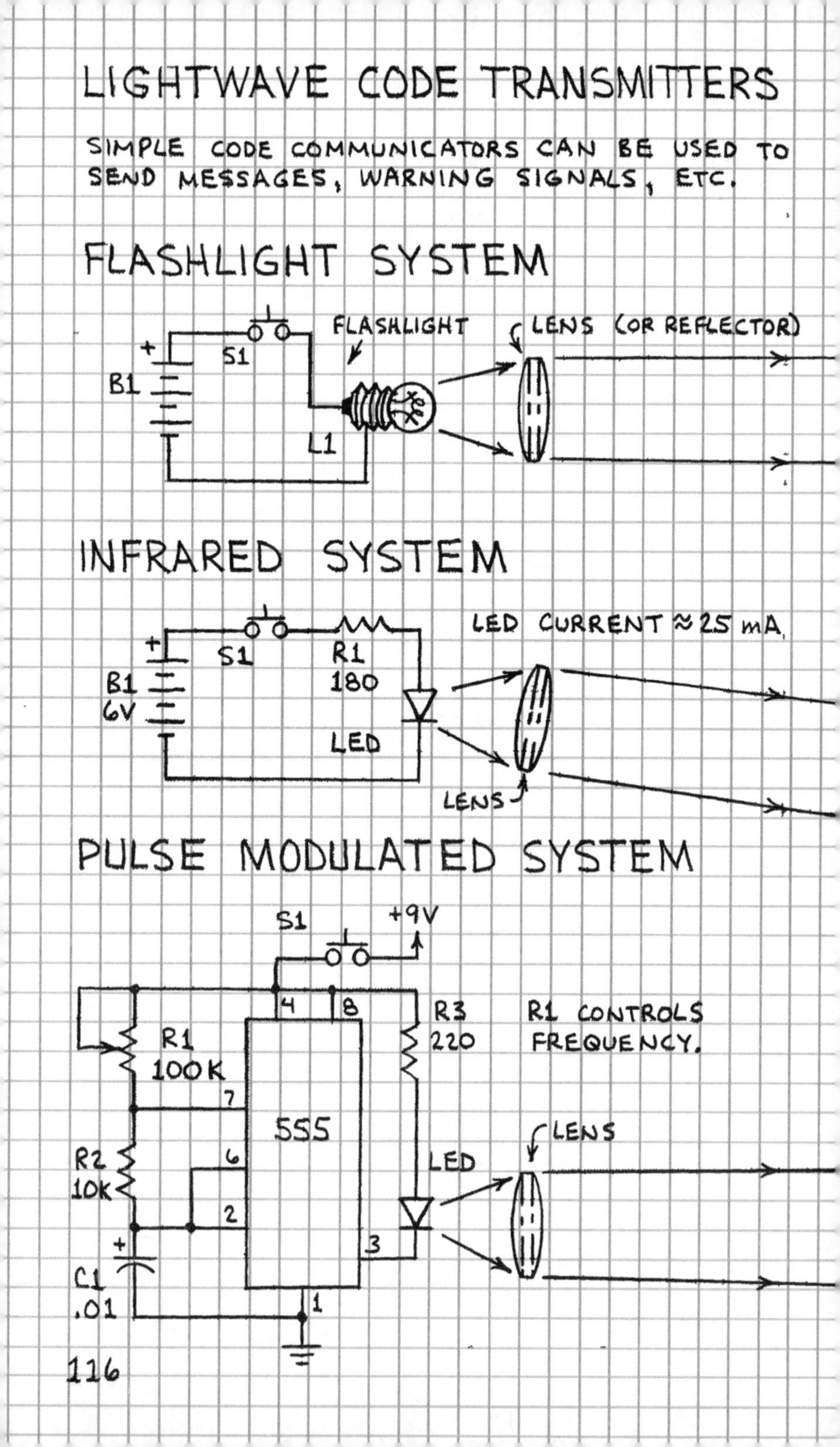
LIGHTWAVE CODE TRANSMITTERS
SIMPLE CODE COMMUNICATORS CAN BE USED TO SEND MESSAGES, WARNING SIGNALS, ETC.
FLASHLIGHT SYSTEM
FLASHLIGHT
LENS (OR REFLECTOR)
+
B1
S1
L1
INFRARED SYSTEM
LED CURRENT ≈ 25 mA.
+
B1
6V
S1
R1
180
LED
LENS
PULSE MODULATED SYSTEM
S1
+9V
4
8
R3
220
R1 CONTROLS FREQUENCY.
R1
100K
7
555
R2
10K
6
2
LENS
LED
+
C1
.01
3
1
116

LIGHTWAVE CODE RECEIVERS

THESE RECEIVERS MUST BE KEPT FROM EXTERNAL LIGHT SOURCES. THE FIRST TWO ARE LIGHT-ACTUATED TONE GENERATORS.

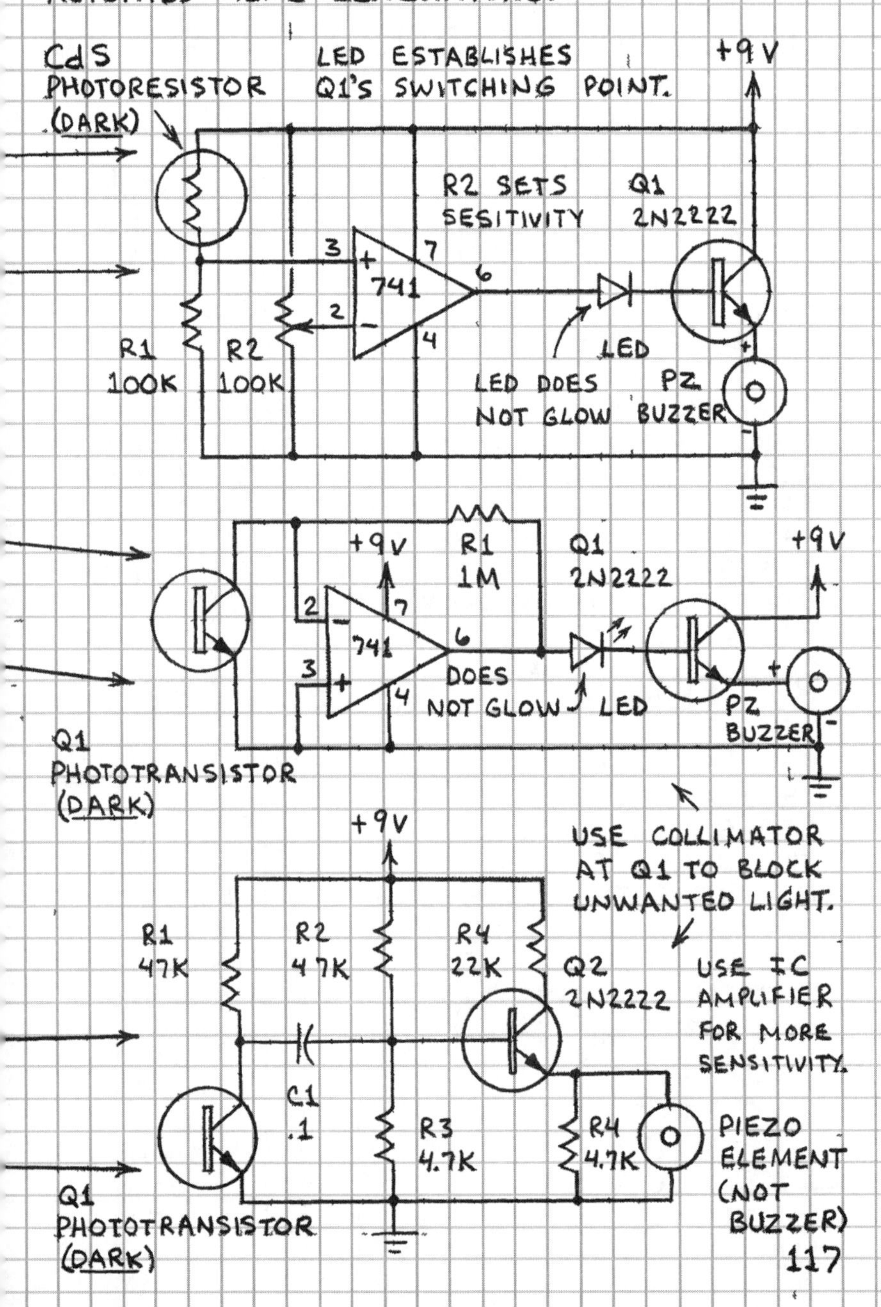

FLASHLIGHT VOICE TRANSMITTERS

THESE SIMPLE AM SYSTEMS DEMONSTRATE THAT INCANDESCENT LAMPS CAN BE VOICE MODULATED.

BASIC VOICE TRANSMITTER

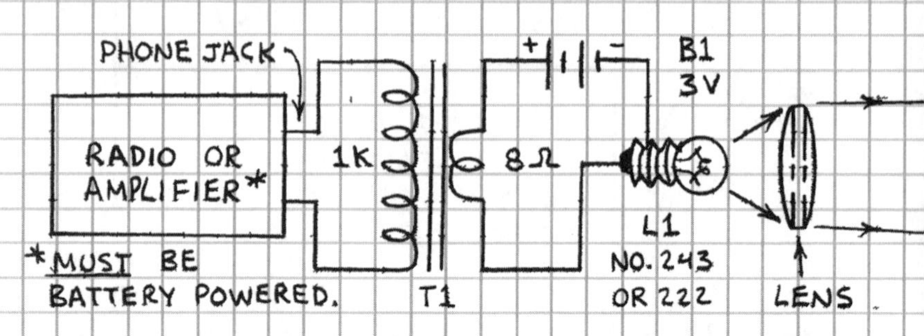

T1 IS MINIATURE 1K:8Ω OUTPUT TRANSFORMER. SINCE MOST PHONE JACKS ARE 8Ω, MUCH BETTER RESULTS WILL BE OBTAINED WITH TWO BACK-TO-BACK TRANSFORMERS. CONNECT 1K WINDINGS OF THE TRANSFORMERS TOGETHER. THEN CONNECT ONE 8Ω WINDING TO RADIO OR AMPLIFIER AND THE OTHER TO THE LAMP AND BATTERY.

BETTER VOICE TRANSMITTER

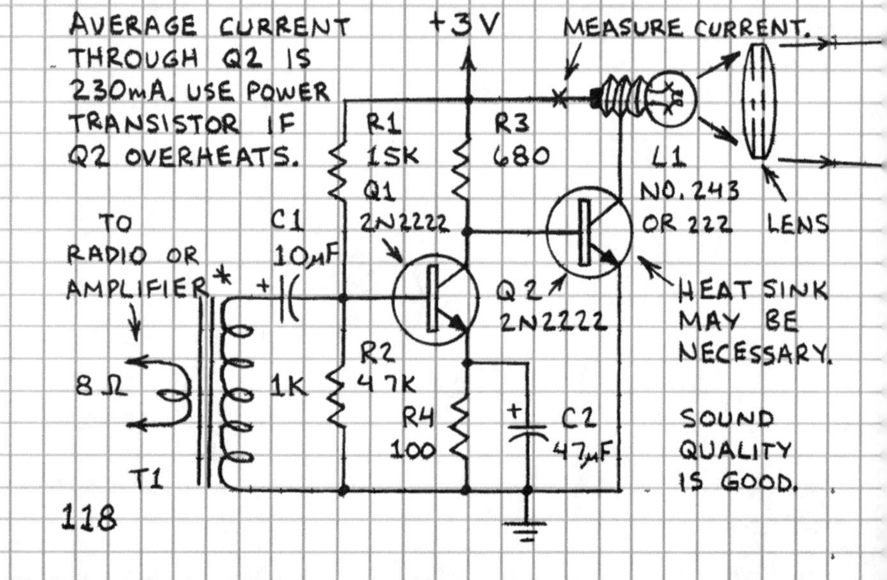

GENERAL PURPOSE RECEIVERS

THESE SIMPLE RECEIVERS CAN RECEIVE ANY AMPLITUDE MODULATED (AM) LIGHTWAVE SIGNALS.

BASIC VOICE RECEIVER

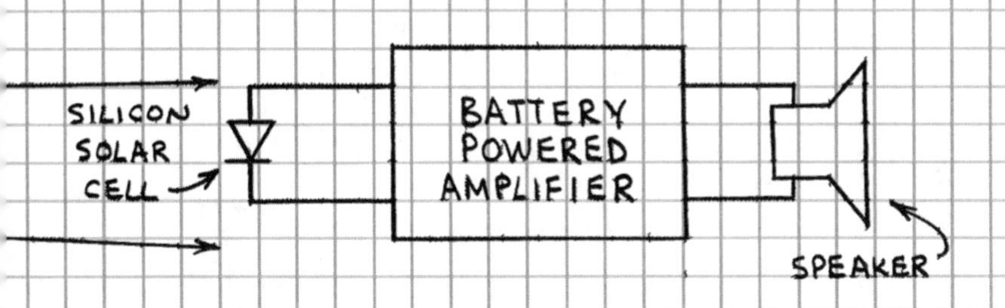

TRANSISTOR VOICE RECEIVER

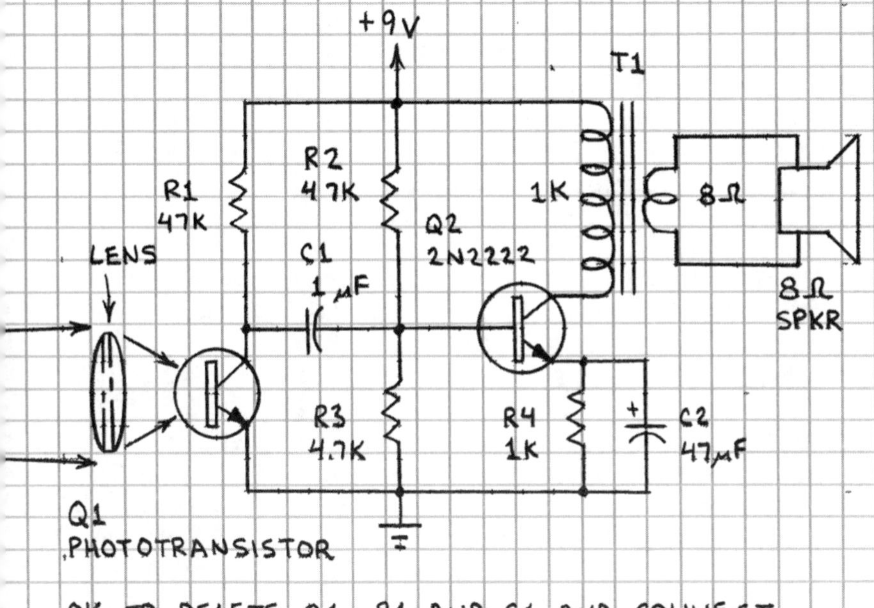

OK TO DELETE Q1, R1 AND C1 AND CONNECT SOLAR CELL BETWEEN Q2'S BASE (CELL ANODE) AND GROUND (CELL CATHODE).

FOR MORE VOLUME USE RECEIVER ON PAGE 121.

AM LIGHTWAVE TRANSMITTER

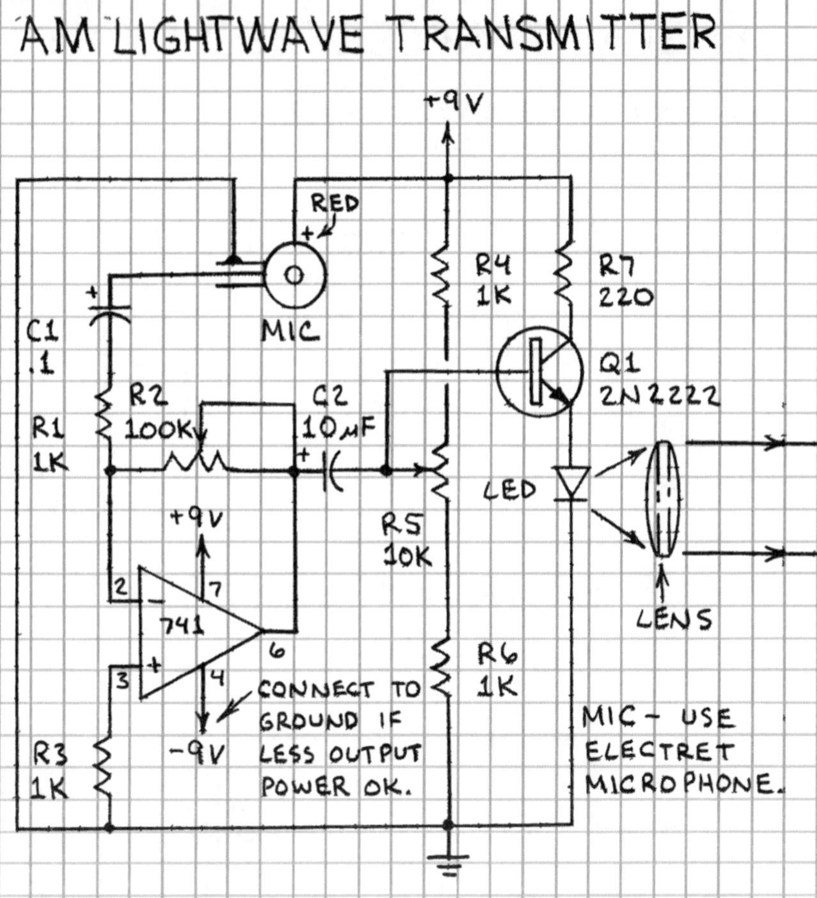

THIS TRANSMITTER WILL SEND YOUR VOICE TO THE RECEIVER ON FACING PAGE. AT NIGHT AND WHEN LENSES ARE USED, A RANGE OF SEVERAL HUNDRED OR MORE FEET IS POSSIBLE. INFRARED LED WILL GIVE BEST RESULTS. HIGH-BRIGHTNESS RED LED WILL ALSO WORK, ESPECIALLY WHEN OPTICAL FIBER IS USED. USE TRIPODS FOR BEST RESULTS IN FREE-SPACE MODE. LENS CAN BE MAGNIFIER.

R2 - GAIN CONTROL
R5 - LED BIAS CONTROL. ADJUST R5 FOR BEST SOUND QUALITY AT RECEIVER.
R7 - LIMITS CURRENT APPLIED TO LED.

KEEP BATTERY LEADS SHORT.

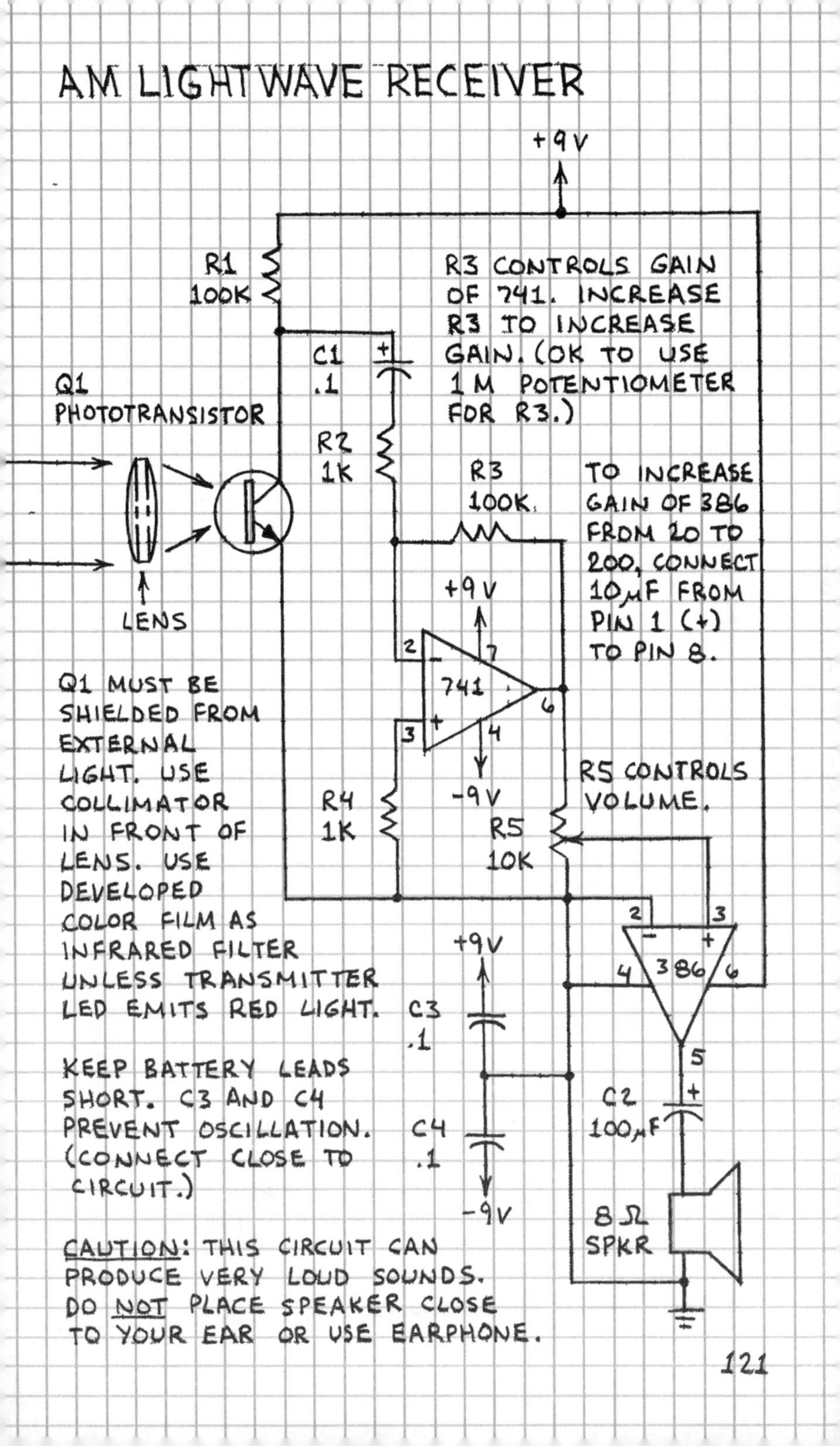
AM LIGHTWAVE RECEIVER
+9V
R1
100K
R3 CONTROLS GAIN OF 741. INCREASE R3 TO INCREASE GAIN. (OK TO USE 1 M POTENTIOMETER FOR R3.)
C1
.1
Q1
PHOTOTRANSISTOR
R2
1K
R3
100K
TO INCREASE GAIN OF 386 FROM 20 TO 200, CONNECT 10µF FROM PIN 1 (+) TO PIN 8.
LENS
+9V
2
7
741
6
3
4
-9V
Q1 MUST BE SHIELDED FROM EXTERNAL LIGHT. USE COLLIMATOR IN FRONT OF LENS. USE DEVELOPED COLOR FILM AS INFRARED FILTER UNLESS TRANSMITTER LED EMITS RED LIGHT.
R4
1K
R5 CONTROLS VOLUME.
R5
10K
2
3
+9V
4
386
6
C3
.1
5
KEEP BATTERY LEADS SHORT. C3 AND C4 PREVENT OSCILLATION. (CONNECT CLOSE TO CIRCUIT.)
C2
100µF
C4
.1
-9V
8 Ω
SPKR
CAUTION: THIS CIRCUIT CAN PRODUCE VERY LOUD SOUNDS. DO NOT PLACE SPEAKER CLOSE TO YOUR EAR OR USE EARPHONE.

PFM LIGHTWAVE TRANSMITTER

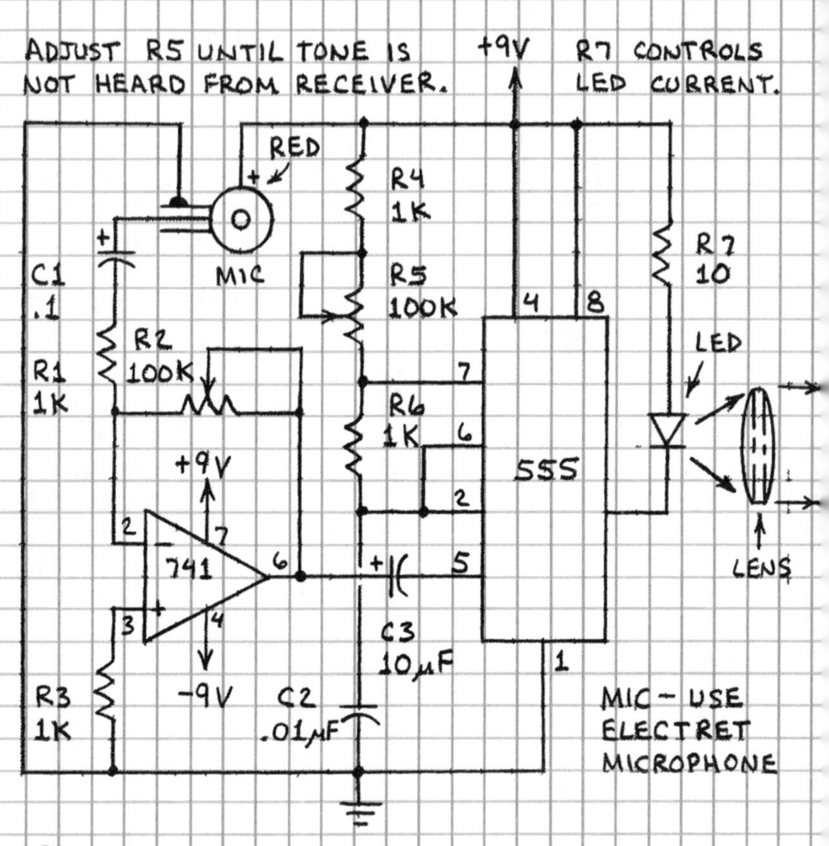

<u>TRANSMITTER:</u> R2 CONTROLS GAIN OF 741 MICROPHONE AMPLIFIER. 555 GENERATES STEADY STREAM OF PULSES HAVING A REPETION RATE CONTROLLED BY R5. AUDIO SIGNAL APPLIED TO PIN 5 OF 555 MODULATES THE PULSE RATE. USE SUPER BRIGHT RED OR INFRARED LED. KEEP BATTERY LEADS SHORT. PFM GIVES UNIFORM RECEIVER VOLUME.

<u>RECEIVER:</u> Q1 RECEIVES PULSES FROM THE LED. THE PULSES ARE AMPLIFIED BY THE FIRST 741. THE SECOND 741 IS CONNECTED AS A COMPARATOR THAT DELIVERS AN OUTPUT PULSE WHEN THE INPUT PULSE EXCEEDS THE REFERENCE VOLTAGE SET BY R4. THE PULSES ARE LOW-PASS FILTERED BY R5 AND C3 AND AMPLIFIED BY THE 386. ADJUST R5 OF TRANSMITTER AND R4 OF RECEIVER FOR BEST SOUND QUALITY.

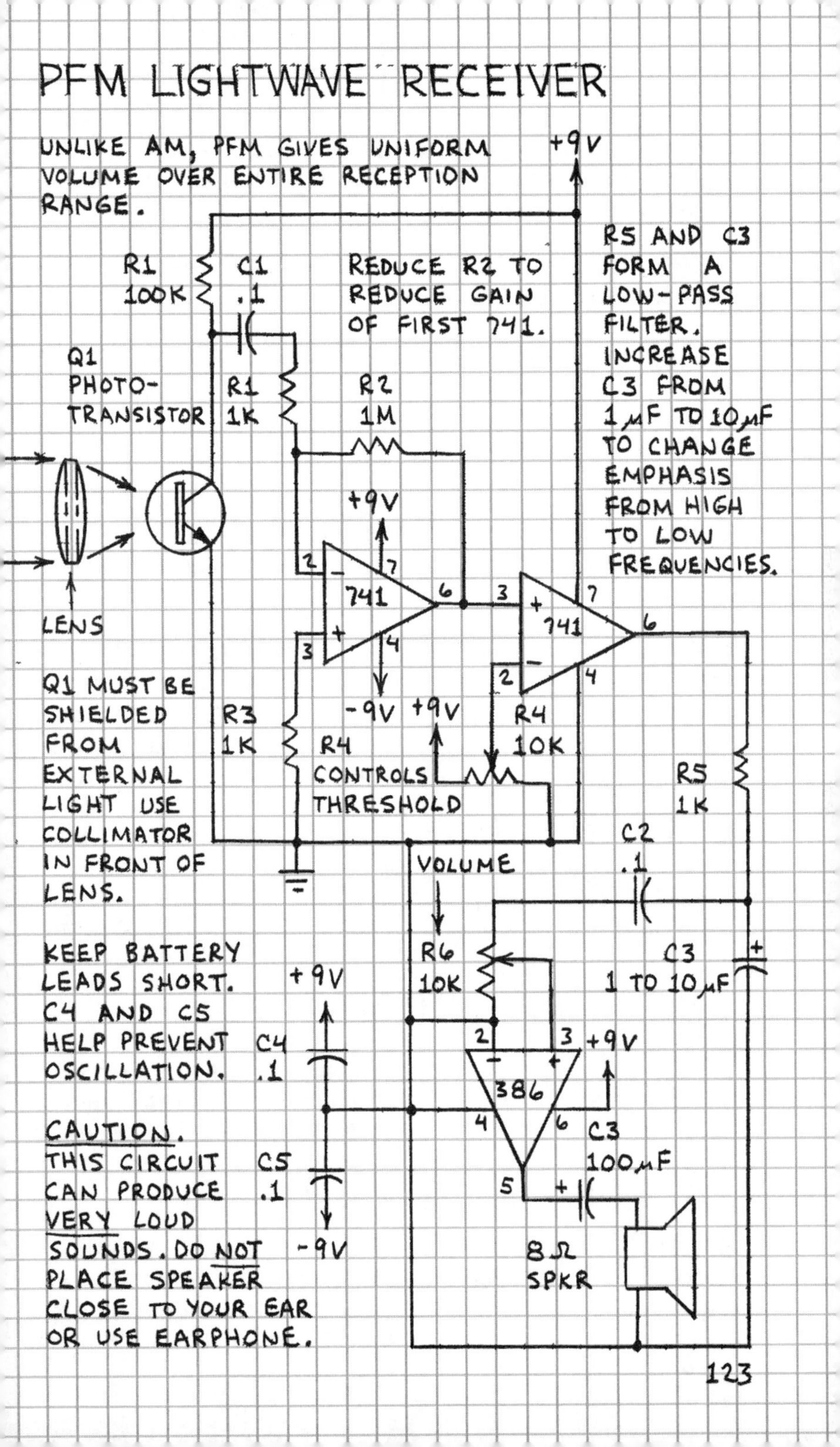
PFM LIGHTWAVE RECEIVER
UNLIKE AM, PFM GIVES UNIFORM VOLUME OVER ENTIRE RECEPTION RANGE.
+9V
R1 100K
C1 .1
REDUCE R2 TO REDUCE GAIN OF FIRST 741.
R5 AND C3 FORM A LOW-PASS FILTER. INCREASE C3 FROM 1µF TO 10µF TO CHANGE EMPHASIS FROM HIGH TO LOW FREQUENCIES.
Q1 PHOTO-TRANSISTOR
R1 1K
R2 1M
+9V
2
7
741
6
3
+
741
7
6
LENS
3
4
2
4
Q1 MUST BE SHIELDED FROM EXTERNAL LIGHT USE COLLIMATOR IN FRONT OF LENS.
R3 1K
-9V +9V
R4 10K
R4 CONTROLS THRESHOLD
R5 1K
C2 .1
VOLUME
R6 10K
C3 1 TO 10µF
KEEP BATTERY LEADS SHORT. C4 AND C5 HELP PREVENT OSCILLATION.
+9V
C4 .1
2
3 +9V
386
4
6
C3 100µF
CAUTION.
THIS CIRCUIT CAN PRODUCE VERY LOUD SOUNDS. DO NOT PLACE SPEAKER CLOSE TO YOUR EAR OR USE EARPHONE.
C5 .1
-9V
5
8Ω SPKR

RADIO COMMUNICATIONS

1886—HEINRICH HERTZ SENT WAVES FROM A SPARK DISCHARGE TO A LOOP OF WIRE. A SMALL SPARK APPEARED AT A GAP IN THE LOOP.

1895—GUGLIELMO MARCONI INVENTED THE WIRELESS TELEGRAPH.

1899— MARCONI SENT "···" ACROSS ATLANTIC OCEAN

MODULATION

WHEN A PURE RADIO-FREQUENCY WAVE (THE CARRIER) IS MIXED WITH A SIGNAL SUCH AS VOICE, THE WAVE IS SAID TO BE MODULATED.

DAMPED WAVE (SPARK GAP)

OK FOR CODE, BUT NOT LEGAL SINCE MANY WAVELENGTHS ARE EMITTED.

CARRIER WAVE

PURE, UNMODULATED RADIO-FREQUENCY WAVE, NO SIGNAL CARRIED.

AMPLITUDE MODULATION

CONSTANT FREQUENCY; AMPLITUDE VARIES WITH INPUT SIGNAL (VOICE, ETC.).

FREQUENCY MODULATION

CONSTANT AMPLITUDE; FREQUENCY VARIES WITH INPUT SIGNAL (VOICE, ETC). GIVES NOISE-FREE RECEPTION.

AMATEUR RADIO

RADIO COMMUNICATION HAS ALWAYS ATTRACTED MANY THOUSANDS OF ENTHUSIASTIC AMATEUR RADIO OPERATORS. THEY WERE AMONG THE FIRST TO DISCOVER THAT SHORTWAVES PERMIT WORLDWIDE COMMUNICATION. THEY PROVIDE COMMUNICATIONS DURING NATURAL DISASTERS AND EMERGENCIES. AND THEY COMMUNICATE WITH FELLOW AMATEURS ACROSS TOWN AND HALFWAY AROUND THE WORLD.

AMATEUR OR HAM RADIO OPERATORS ARE LICENSED AND ASSIGNED A CALL SIGN BY THE FEDERAL GOVERNMENT. PROSPECTIVE HAMS MUST PASS A WRITTEN EXAM. FOR MORE INFORMATION, CONTACT THE AMERICAN RADIO RELAY LEAGUE (ARRL) IN NEWINGTON, CT 06111. THE ARRL SELLS EXCELLENT PUBLICATIONS FOR BOTH PROSPECTIVE AND ESTABLISHED HAMS.

CITIZENS BAND RADIO

THE CITIZENS BAND IS 40 CHANNELS IN THE VICINITY OF 27 MHz. THESE CHANNELS ARE INTENDED FOR TWO-WAY PERSONAL AND BUSINESS COMMUNICATION. ONE CHANNEL (9) IS RESERVED FOR EMERGENCY TRANSMISSIONS. THOUGH NO LICENSE IS REQUIRED, CITIZENS BAND (CB) OPERATORS HAVE FEWER PRIVILEGES THAN AMATEUR RADIO OPERATORS. FOR EXAMPLE, MAXIMUM TRANSMITTED POWER IS LIMITED TO 4 WATTS.

FEDERAL COMMUNICATIONS COMMISSION

THE FEDERAL COMMUNICATIONS COMMISSION (FCC) REGULATES RADIO COMMUNICATION IN THE UNITED STATES. VIOLATIONS OF FCC REGULATIONS CAN RESULT IN SEVERE PENALTIES. YOU CAN WRITE THE FCC (GETTYSBURG, PA 17326) TO REQUEST INFORMATION ABOUT ITS REGULATIONS.

DIODE RECEIVER BASICS

A RADIO-FREQUENCY (RF) ELECTROMAGNETIC WAVE WILL CAUSE A FLUCTUATING CURRENT TO FLOW IN A WIRE ANTENNA:

CURRENT PRODUCED BY TONE-MODULATED RF SIGNAL.

CURRENT PRODUCED BY VOICE-MODULATED RF SIGNAL.

THE FLUCTUATING CURRENT CAN BE TRANSFORMED INTO SOUND BY REMOVING THE POSITIVE OR NEGATIVE HALF OF THE WAVE WITH A DIODE:

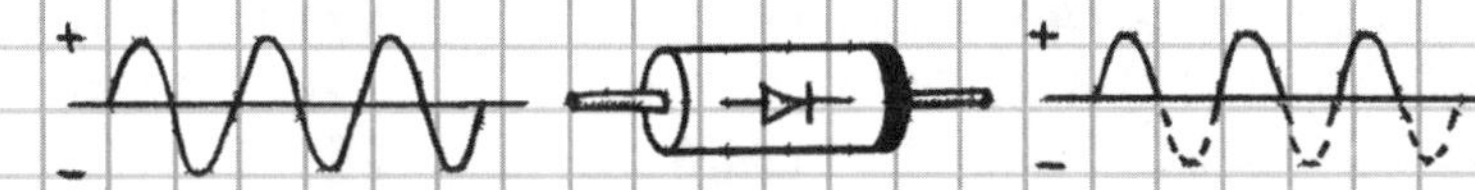

THE SIGNAL IS NOW SAID TO BE RECTIFIED. THE TWO HALVES OF THE WAVE WILL NOT CANCEL ONE ANOTHER WHEN THE OUTPUT IS MONITORED. THEREFORE THE AUDIO SIGNAL SUPERIMPOSED ON THE RF SIGNAL CAN BE HEARD FROM A SMALL EARPHONE CONNECTED TO THE DIODE.

SIMPLE RF TUNING COIL

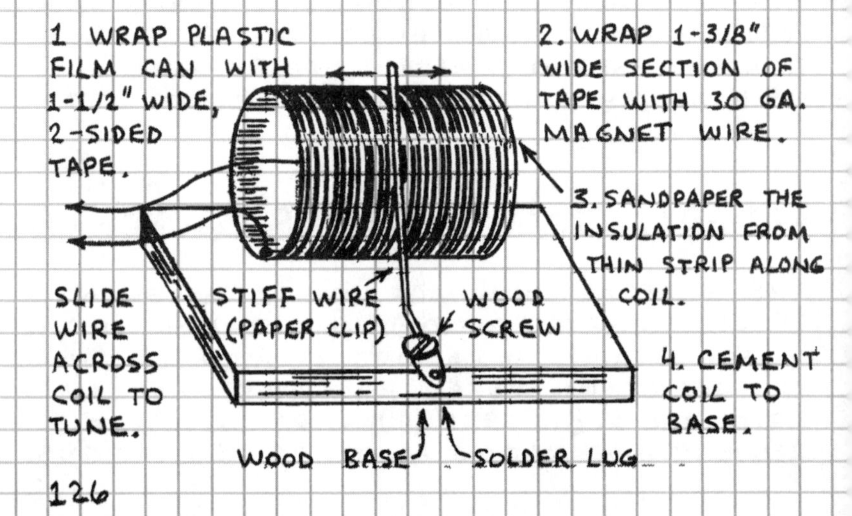

SIMPLE DIODE RECEIVER

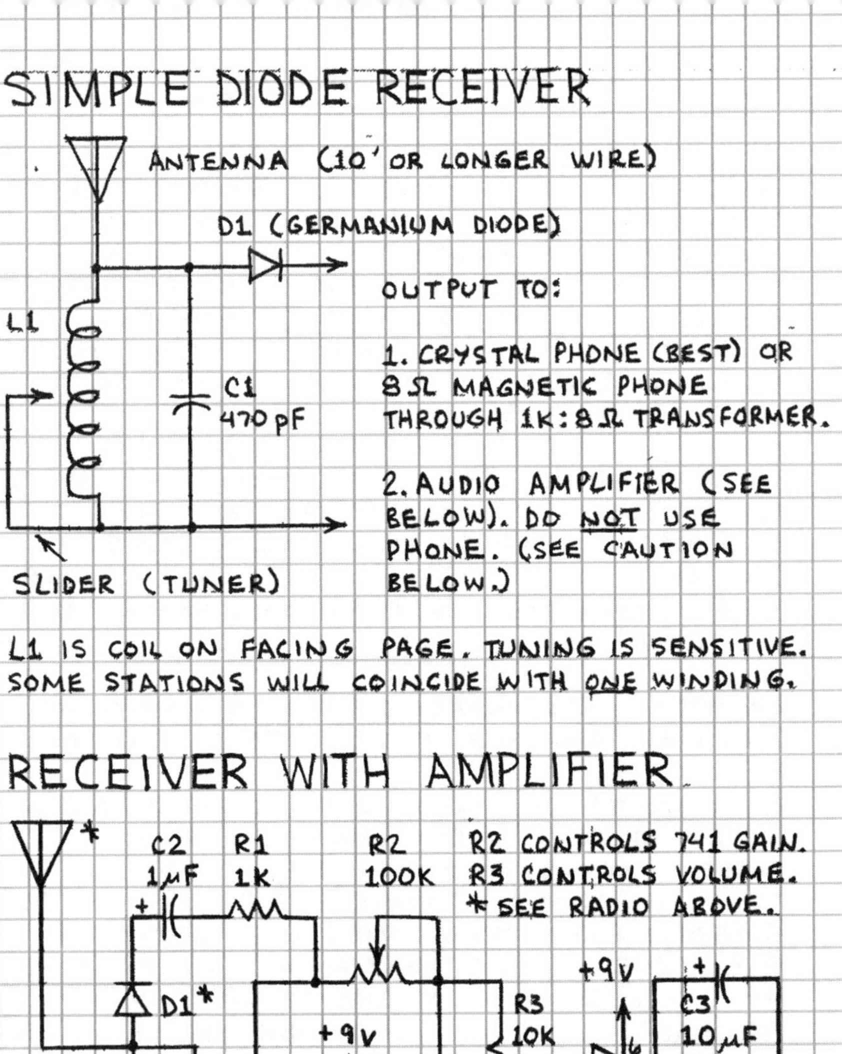

OUTPUT TO:

1. CRYSTAL PHONE (BEST) OR 8 Ω MAGNETIC PHONE THROUGH 1K:8 Ω TRANSFORMER.

2. AUDIO AMPLIFIER (SEE BELOW). DO <u>NOT</u> USE PHONE. (SEE CAUTION BELOW.)

L1 IS COIL ON FACING PAGE. TUNING IS SENSITIVE. SOME STATIONS WILL COINCIDE WITH <u>ONE</u> WINDING.

RECEIVER WITH AMPLIFIER

R2 CONTROLS 741 GAIN.
R3 CONTROLS VOLUME.
* SEE RADIO ABOVE.

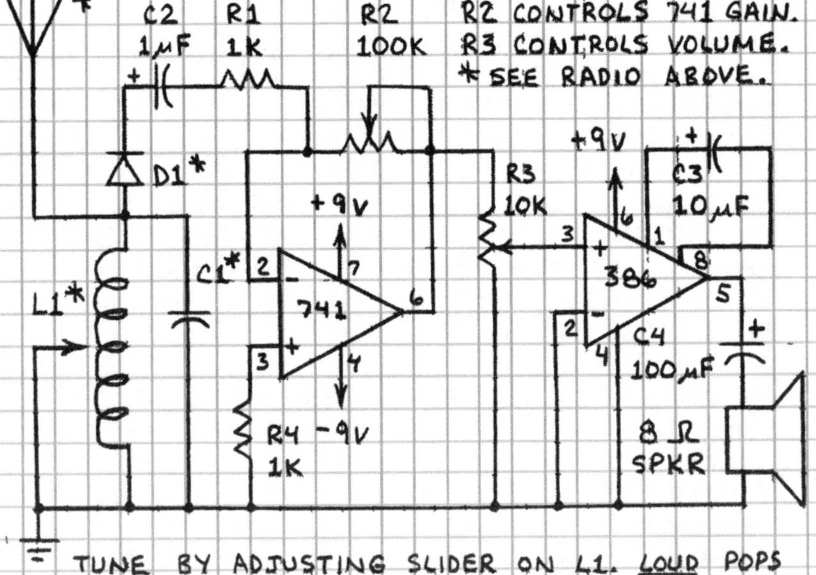

TUNE BY ADJUSTING SLIDER ON L1. <u>LOUD</u> POPS MAY OCCUR WHEN SLIDER IS MOVED. VOLUME CAN BE <u>VERY</u> LOUD. <u>CAUTION</u>: DON'T USE EARPHONES!

SHORTWAVE LISTENING

FEW HOBBIES ARE AS REWARDING OR INTELLECTUALLY STIMULATING AS SHORTWAVE LISTENING. YET MANY PEOPLE HAVE NEVER LISTENED TO A SHORTWAVE RADIO. EVEN A VERY INEXPENSIVE SHORTWAVE RADIO CAN RECEIVE BROADCASTS FROM HUNDREDS OF STATIONS AROUND THE WORLD. MANY OF THEM ARE IN ENGLISH. SHORTWAVE BROADCASTS CAN BE DIVIDED INTO THREE BROAD CATEGORIES:

INTERNATIONAL BROADCASTS—THESE ORIGINATE FROM BOTH PRIVATE AND GOVERNMENT STATIONS AND ARE INTENDED FOR A WIDE AUDIENCE. PROGRAMMING, OFTEN IN ENGLISH, INCLUDES NEWS, WEATHER, INTERVIEWS, DRAMA AND LISTENER MAIL.

PERSONAL COMMUNICATIONS—THIS CATEGORY INCLUDES AMATEUR AND CITIZENS BAND RADIO.

UTILITIES—VIRTUALLY ALL BROADCASTS NOT LISTED ABOVE CAN BE CONSIDERED UTILITIES. THESE INCLUDE TIME SIGNALS, COMPUTER TRANSMISSIONS, WEATHER REPORTS, SATELLITE SIGNALS AND MANY KINDS OF INDUSTRIAL AND GOVERNMENT TRANSMISSIONS. INCLUDED ARE TELECOMMUNICATIONS TO AND FROM SHIPS, AIRCRAFT, TAXIS AND COMMERCIAL VEHICLES. ALSO INCLUDED ARE TRANSMISSIONS FROM SPY, RADIO CONTROL, TRACKING, SURVEILLANCE, TELEMETRY, WEATHER BALLOON AND OCEAN BUOY TRANSMITTERS.

MANY OF THESE TRANSMISSIONS ARE BROADCAST AT FREQUENCIES BETWEEN THE BROADCAST BAND AND 30 MHz. THE SIMPLE RECEIVER ON THE FACING PAGE CAN RECEIVE SIGNALS FROM 1 TO 6 MHz. IN ONE EVENING THIS RADIO RECEIVED SIGNALS FROM ASIA, EUROPE, SOUTH AMERICA AND NORTH AMERICA. THE ANTENNA WAS A 14' INDOOR WIRE.

SHORTWAVE RECEIVER

THIS SIMPLE RECEIVER CAN BE ASSEMBLED ON A SOLDERLESS BREADBOARD. THOUGH THIS RECEIVER DOES NOT SEPARATE STATIONS AS WELL AS A COMMERCIAL RECEIVER, IT IS SURPRISINGLY SENSITIVE AND WILL RECEIVE STATIONS FROM AROUND THE WORLD.

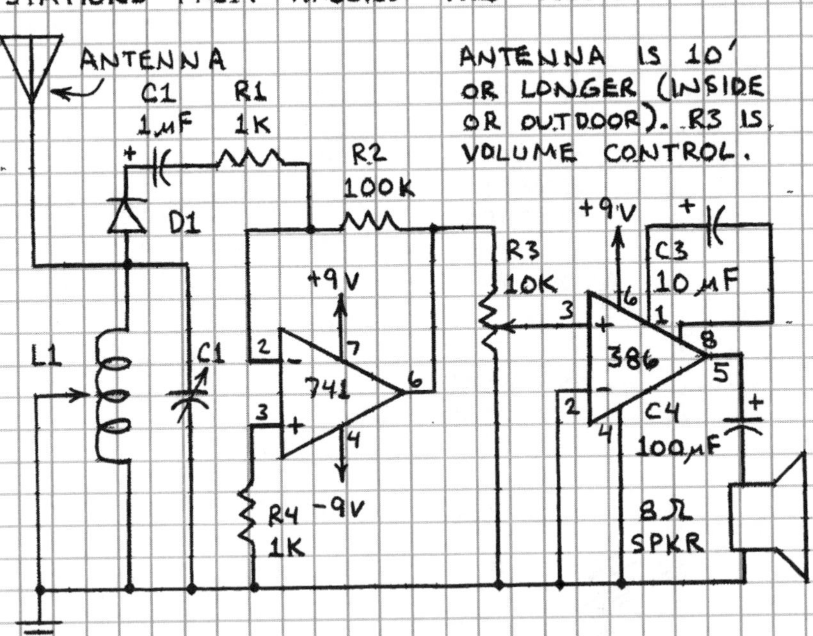

ANTENNA IS 10′ OR LONGER (INSIDE OR OUTDOOR). R3 IS VOLUME CONTROL.

L1 IS 25-50 TURNS OF 30 GAUGE MAGNET WIRE WRAPPED AROUND PLASTIC FILM CAN. SEE TUNING COIL ASSEMBLY DETAILS ON PAGE 34.

C1 IS 10-365 pF VARIABLE CAPACITOR FROM DISCARDED RADIO OR 10-40 pF OR SO CRYSTAL OSCILLATOR TUNING CAPACITOR.

TUNE BY SETTING L1'S SLIDER TO ANY POSITION AND ADJUST C1. CHANGE L1'S SLIDER POSITION FOR DIFFERENT FREQUENCY RANGES.

<u>CAUTION:</u> VOLUME CAN BE <u>VERY</u> LOUD, ESPECIALLY WHEN L1'S SLIDER IS MOVED AWAY FROM L1 AND LOCAL STATIONS BOOM IN. <u>NO</u> EARPHONES!

ANTENNAS

THE PERFORMANCE OF RADIO TRANSMITTERS AND RECEIVERS IS VERY MUCH DEPENDENT ON THEIR ANTENNAS. THE SIMPLEST ANTENNA IS A WIRE OR ROD WHOSE LENGTH EQUALS OR IS 1/4 OR 1/2 THE WAVELENGTH OF THE RECEIVED SIGNAL. THREE COMMON WIRE ANTENNAS ARE:

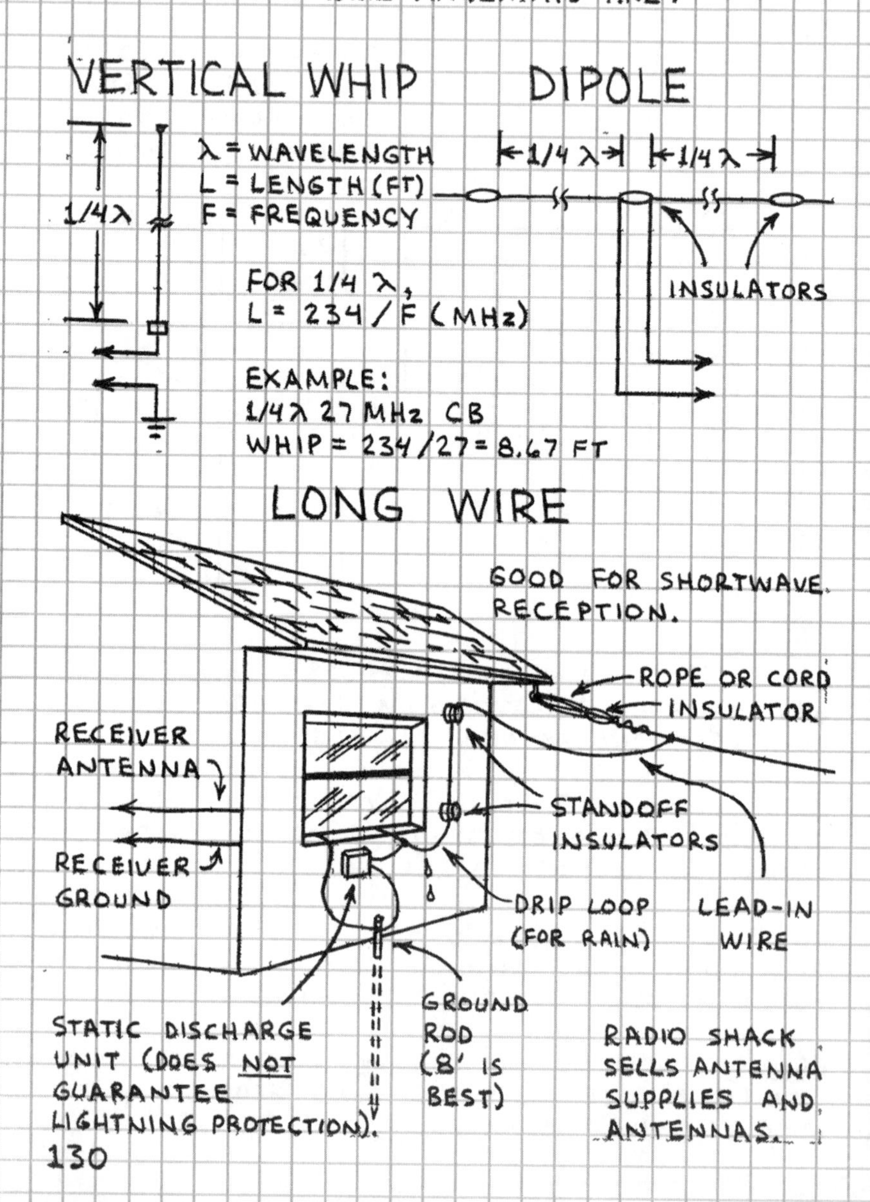

ANTENNA SAFETY

THE INSTALLATION OF AN ANTENNA REQUIRES CAREFUL ATTENTION TO SAFETY. CARELESSNESS CAN RESULT IN SERIOUS INJURY OR A FATAL ELECTRICAL SHOCK. YOU MUST:

1. <u>NEVER</u> INSTALL ANY PART OF AN ANTENNA NEAR A POWER LINE.

2. <u>NEVER</u> TOUCH ANY PART OF AN ANTENNA THAT CONTACTS A POWER LINE.

3. <u>DISCONNECT</u> AND <u>DO NOT USE</u> AN ANTENNA DURING AN ELECTRICAL STORM.

4. CONNECT OUTDOOR ANTENNAS TO A WELL GROUNDED STATIC DISCHARGE UNIT.

5. READ THE ANTENNA SAFETY TIPS SUPPLIED WITH COMMERCIAL ANTENNAS AND GIVEN IN "THE ARRL ANTENNA HANDBOOK."

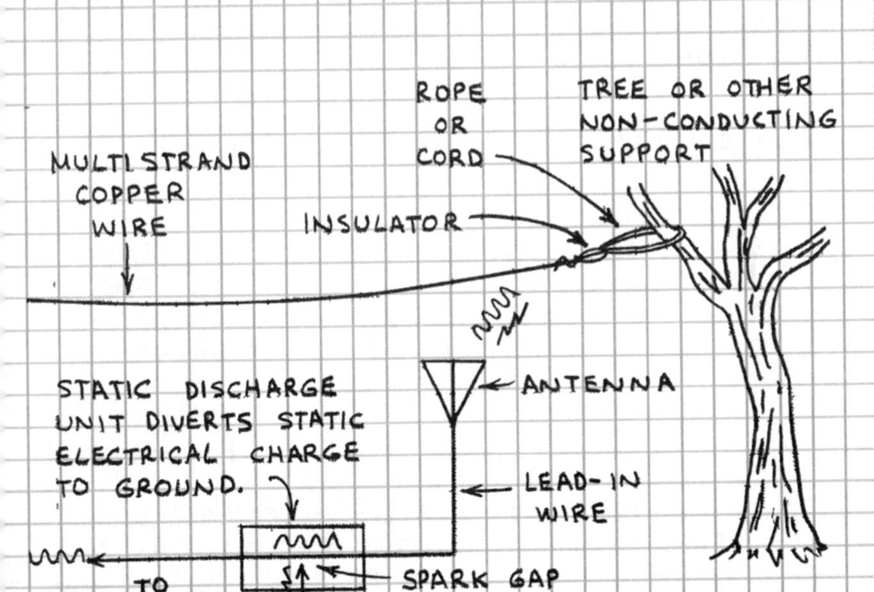

BASIC RADIO TRANSMITTERS

RADIO-FREQUENCY (RF) WAVES ARE CREATED WHEN AN ELECTRICAL CURRENT IS SWITCHED RAPIDLY ON OR OFF. THIS IS WHY A RADIO RECEIVER EMITS A BURST OF STATIC DURING A LIGHTNING DISCHARGE OR A POP WHEN A NEARBY APPLIANCE IS SWITCHED ON.

BROADBAND RF TRANSMITTER

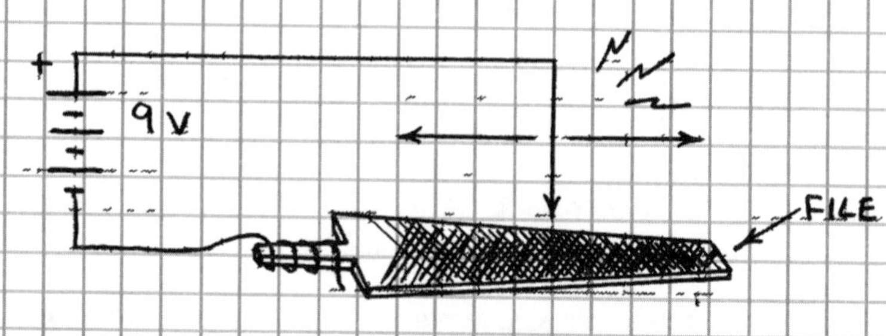

STROKE WIRE ACROSS FILE. BURSTS OF NOISE WILL BE EMITTED BY A NEARBY RADIO. SINCE MANY DIFFERENT WAVELENGTHS ARE PRODUCED ("HASH"), THE SIGNAL IS EQUALLY STRONG ACROSS THE BROADCAST BAND.

BROADBAND PULSE TRANSMITTER

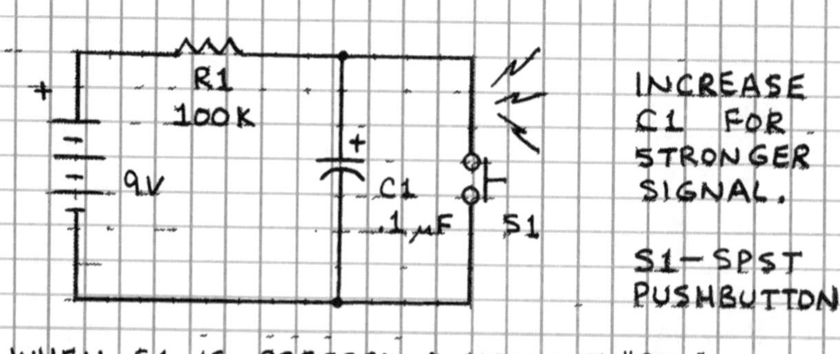

INCREASE C1 FOR STRONGER SIGNAL.

S1 — SPST PUSHBUTTON

WHEN S1 IS PRESSED A DISTINCT "POP" WILL BE HEARD FROM A NEARBY RADIO. THIS CIRCUIT AVOIDS A DIRECT SHORT CIRCUIT ACROSS THE BATTERY. INSTEAD C1 IS SHORTED BY S1 AFTER BEING CHARGED THROUGH R1.

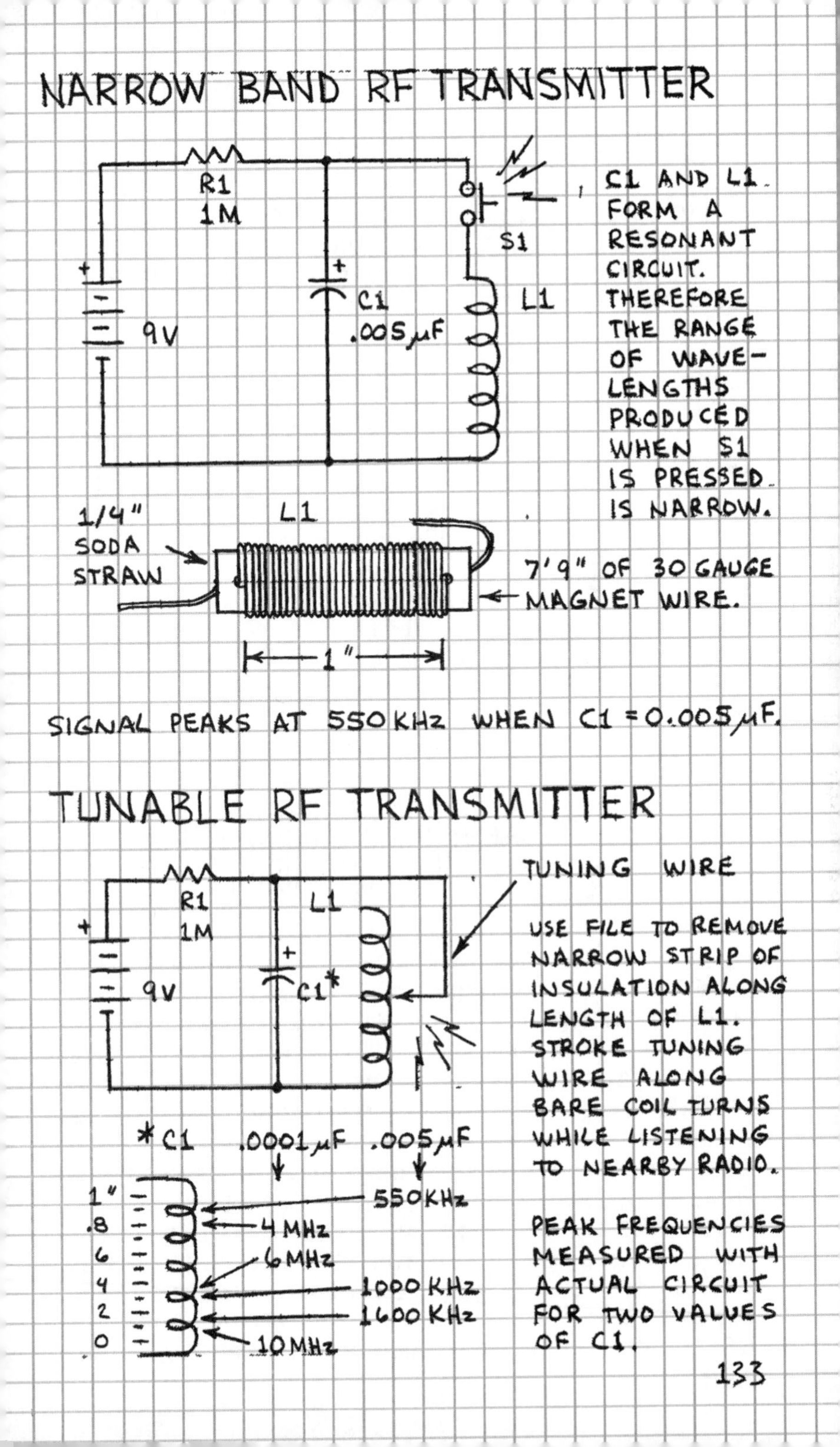
NARROW BAND RF TRANSMITTER
R1
1M
S1
+
9V
+
C1
.005 μF
L1
C1 AND L1 FORM A RESONANT CIRCUIT. THEREFORE THE RANGE OF WAVE-LENGTHS PRODUCED WHEN S1 IS PRESSED IS NARROW.
1/4" SODA STRAW
L1
7'9" OF 30 GAUGE MAGNET WIRE.
1"
SIGNAL PEAKS AT 550 KHz WHEN C1 = 0.005 μF.
TUNABLE RF TRANSMITTER
R1
1M
L1
+
9V
+
C1*
TUNING WIRE
USE FILE TO REMOVE NARROW STRIP OF INSULATION ALONG LENGTH OF L1. STROKE TUNING WIRE ALONG BARE COIL TURNS WHILE LISTENING TO NEARBY RADIO.
*C1
.0001 μF
.005 μF
1"
.8
6
4
2
0
550 KHz
4 MHz
6 MHz
1000 KHz
1600 KHz
10 MHz
PEAK FREQUENCIES MEASURED WITH ACTUAL CIRCUIT FOR TWO VALUES OF C1.
133

TRANSISTOR RF TRANSMITTER

A SINGLE TRANSISTOR CAN BE CONNECTED AS AN OSCILLATOR THAT SUPPLIES A SERIES OF RADIO-FREQUENCY PULSES. THE BASIC HARTLEY OSCILLATOR SHOWN HERE WILL SEND RF PULSES TO A SHORTWAVE OR BROADCAST BAND RADIO SEVERAL FEET AWAY.

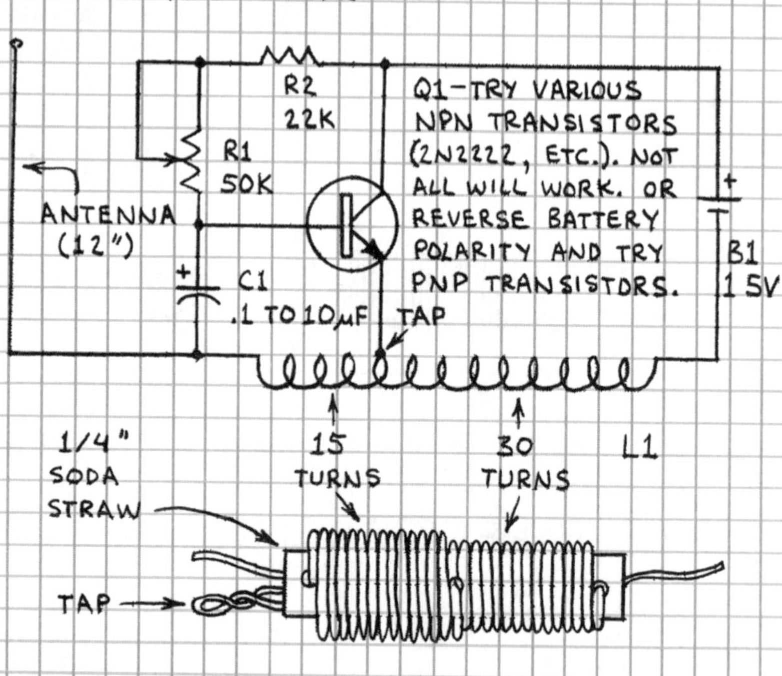

L1 IS A HOMEMADE AIR-CORE RF COIL. USE 30 GAUGE WRAPPING WIRE OR MAGNET WIRE. (USE MAGNET WIRE FOR SMALLER COIL. BURN THE VARNISH FROM ENDS OF L1 WITH A MATCH AND LIGHTLY BUFF CHARRED VARNISH WITH SAND PAPER.) BEFORE WINDING, PUNCH SMALL HOLE IN ONE END OF STRAW (RIGHT END OF COIL ABOVE). INSERT 2" OF WIRE THROUGH HOLE AND WIND 30 TURNS. PUNCH SECOND SMALL HOLE (LEFT END OF COIL) AND INSERT 2" LOOP OF WIRE (TAP) THROUGH HOLE. WIND BACK 15 TURNS BACK OVER FIRST WINDING, PUNCH HOLE THROUGH WINDING AND INSERT END OF WIRE. IF WRAPPING WIRE IS USED, CUT TAP LOOP AND TWIST EXPOSED WIRES.

C1: USE 0.1 μF TO TRANSMIT AN AUDIO TONE. USE 10 μF TO TRANSMIT A STREAM OF POCKS. USE A MINIATURE ELECTROLYTIC CAPACITOR.

R1: CHANGE R1'S SETTING TO VARY OSCILLATION FREQUENCY.

B1: USE A PENLIGHT CELL OR A MERCURY OR SILVER OXIDE BUTTON CELL. <u>WARNING</u>: <u>NEVER</u> ATTEMPT TO SOLDER LEADS TO MINIATURE POWER CELLS! THEY <u>WILL</u> EXPLODE.

CIRCUIT OPERATION

THIS TRANSMITTER EMITS AN RF SIGNAL THAT CAN BE RECEIVED ACROSS A WIDE PART OF THE BROADCAST AND SHORTWAVE SPECTRUM, PARTICULARLY THE 16-METER BAND AND BEYOND. THE SIGNAL CAN ALSO BE RECEIVED AT THE LOW END OF THE 88- TO 108-MHz FM BAND.

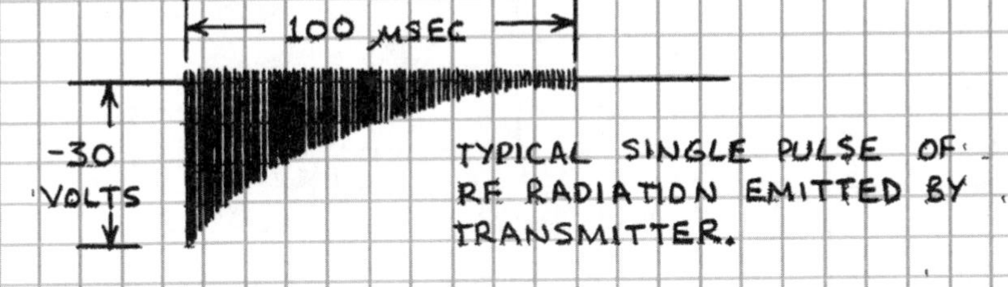

TYPICAL SINGLE PULSE OF RF RADIATION EMITTED BY TRANSMITTER.

EACH TRANSMITTED PULSE IS AN ENVELOPE OF BROAD SPECTRUM RF OSCILLATIONS RATHER THAN A PURE, SINGLE FREQUENCY SIGNAL. NOTE THAT THE AUTOTRANSFORMER ACTION OF L1 INCREASES THE OUTPUT FROM 1.5 TO -30 VOLTS.

TO TRANSMIT TEMPERATURE OR LIGHT INTENSITY, REPLACE R1 WITH A THERMISTOR OR CADMIUM SULFIDE PHOTORESISTOR. USE A VALUE FOR C1 THAT GIVES A PULSE RATE OF A FEW PULSES PER SECOND. WITH THE HELP OF A DIGITAL WATCH OR TIMER, YOU CAN THEN COUNT THE NUMBER OF PULSES IN, SAY, 10 SECONDS FOR EACH OF SEVERAL INPUT CONDITIONS.

CODE TRANSMITTER

THIS TRANSMITTER WILL SEND TONE TO NEARBY BROADCAST BAND RADIO TUNED TO NEAR 700 KHz. TRANSMITTING RANGE IS SEVERAL FEET.

L1 IS AIR CORE COIL. USE 8' OF 30 GAUGE MAGNET WIRE. TAP IS AT CENTER OF COIL.

PRESS S1 TO TRANSMIT TONE.

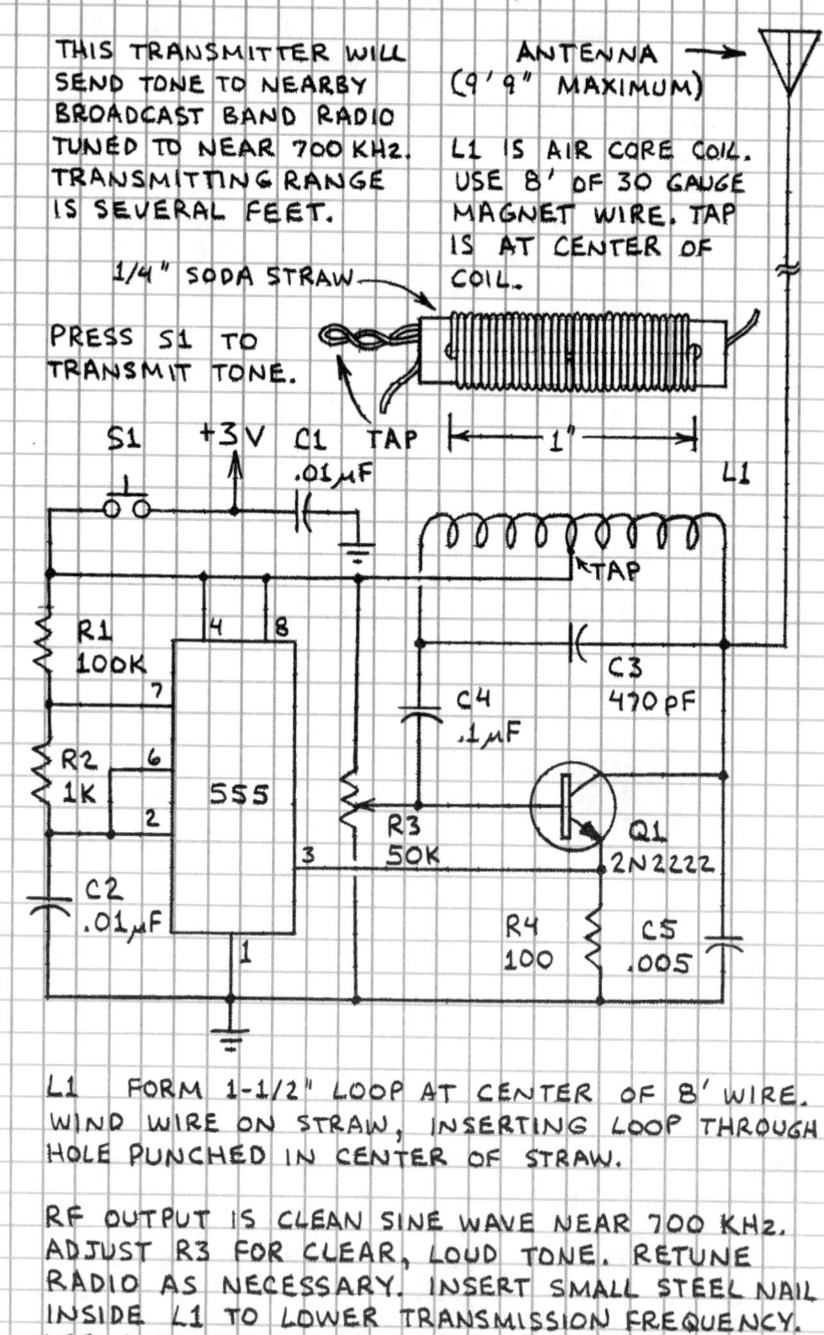

L1 FORM 1-1/2" LOOP AT CENTER OF 8' WIRE. WIND WIRE ON STRAW, INSERTING LOOP THROUGH HOLE PUNCHED IN CENTER OF STRAW.

RF OUTPUT IS CLEAN SINE WAVE NEAR 700 KHz. ADJUST R3 FOR CLEAR, LOUD TONE. RETUNE RADIO AS NECESSARY. INSERT SMALL STEEL NAIL INSIDE L1 TO LOWER TRANSMISSION FREQUENCY. USE DURING DAY FOR MAXIMUM RANGE.

VOICE TRANSMITTER

THE RF OSCILLATOR OF THIS TRANSMITTER IS IDENTICAL TO THE ONE ON THE FACING PAGE. REFER THERE FOR L1 ASSEMBLY.

INPUT IS ELECTRET MICROPHONE. OK TO CONNECT 1K SIDE OF 8Ω : 1K TRANSFORMER TO INPUT.

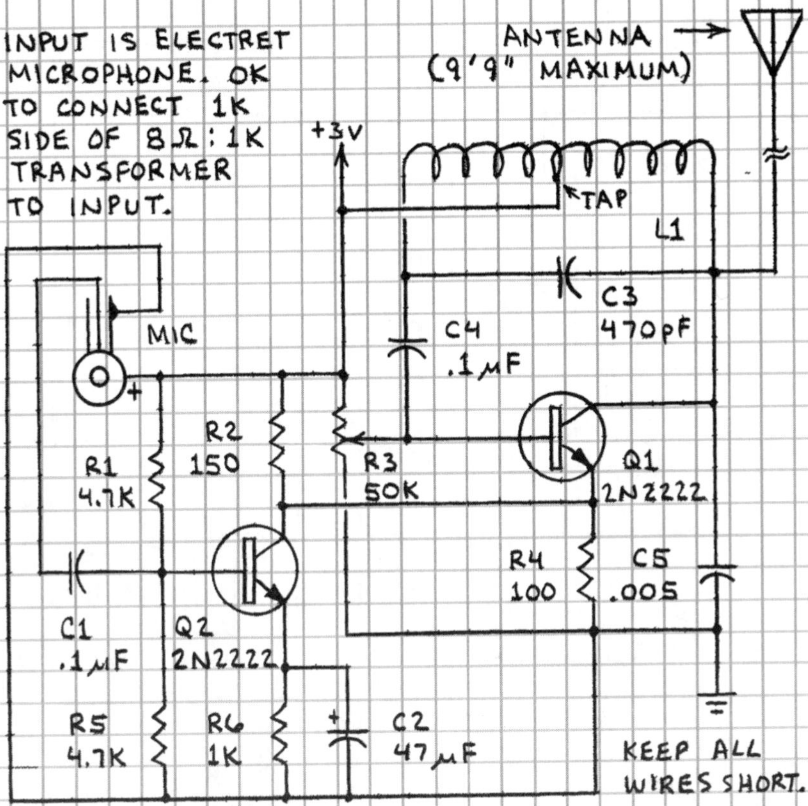

ANTENNA (9'9" MAXIMUM)

+3V

TAP

L1

C3
470pF

MIC

C4
.1μF

R2
150

R1
4.7K

R3
50K

Q1
2N2222

R4
100

C5
.005

C1
.1μF

Q2
2N2222

R5
4.7K

R6
1K

C2
47μF

KEEP ALL WIRES SHORT.

RF OUTPUT IS CLEAN SINE WAVE NEAR 700 KHz. PLACE MICROPHONE CLOSE TO EARPHONE CONNECTED TO TAPE RECORDER. THEN TUNE NEARBY RADIO TO RECEIVE SIGNAL FROM TRANSMITTER. ADJUST R3 FOR BEST SOUND. RETUNE RADIO AS NECESSARY. REMOVE RECORDER AND SPEAK INTO MICROPHONE.

THE TRANSMITTERS ON THIS AND FACING PAGE CONFORM TO THE REQUIREMENTS OF THE FCC GIVEN IN 47 CFR, PART 15.113 WHEN R3 IS ADJUSTED FOR CLEAREST OUTPUT SIGNAL, B1 IS 3 VOLTS AND THE ANTENNA LENGTH < 3 METERS.

AUTOMATIC TONE TRANSMITTER

THIS CIRCUIT TRANSMITS A BRIEF (1/4 SECOND) TONE BURST ONCE EVERY 10 SECONDS TO AN FM BAND RECEIVER UP TO A FEW HUNDRED FEET AWAY.

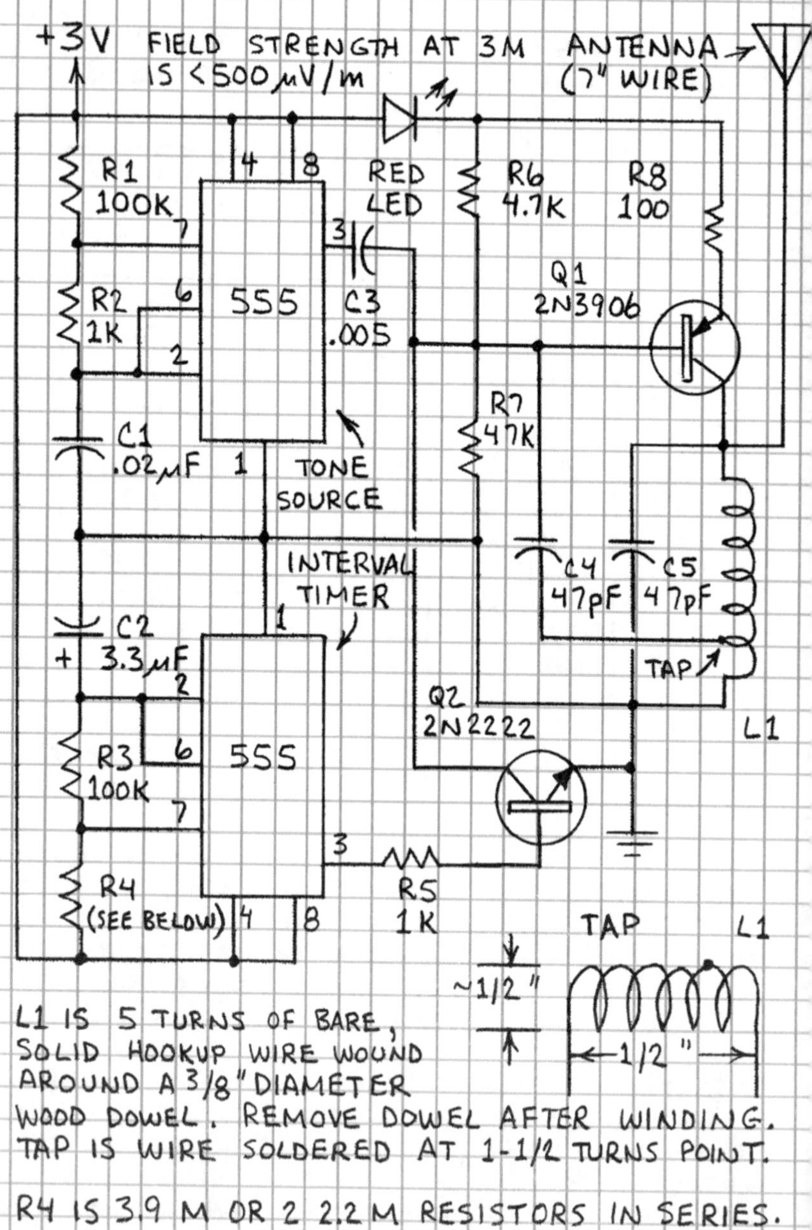

L1 IS 5 TURNS OF BARE, SOLID HOOKUP WIRE WOUND AROUND A 3/8" DIAMETER WOOD DOWEL. REMOVE DOWEL AFTER WINDING. TAP IS WIRE SOLDERED AT 1-1/2 TURNS POINT.

R4 IS 3.9 M OR 2 2.2 M RESISTORS IN SERIES.

CIRCUIT OPERATION

Q1 OSCILLATES AT A FREQUENCY CONTROLLED BY C5 AND L1. VALUES SHOWN GIVE FREQUENCY NEAR 100 MHz. USE VARIABLE CAPACITOR FOR C5 TO VARY FREQUENCY.

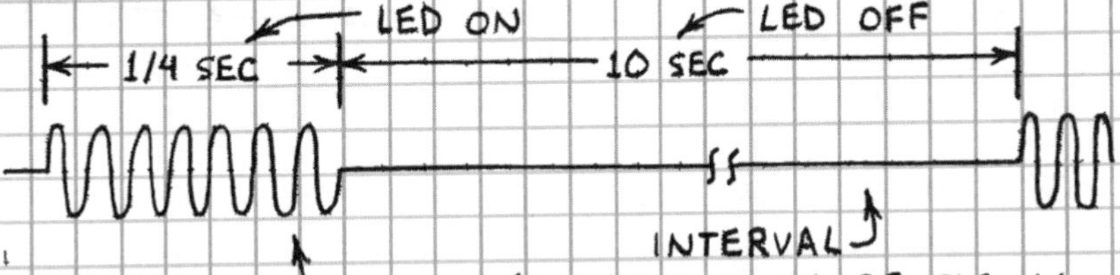

LED ON
LED OFF
1/4 SEC
10 SEC
INTERVAL
RF SIGNAL
(AUDIO FREQUENCY CONTROLLED BY R1/C1)
(DURATIONS OF RF SIGNAL AND INTERVAL CONTROLLED BY R4/R3/C2)

TO ADJUST, DISCONNECT Q2'S COLLECTOR FROM C3. TUNE FM RADIO UNTIL STEADY TONE IS RECEIVED. RECONNECT Q2. DO NOT OPERATE CIRCUIT IN CONTINUOUS TONE MODE UNLESS ADJUSTMENTS ARE BEING MADE. (SEE FCC RULES BELOW AND ON FOLLOWING PAGE.) INSTALL CIRCUIT IN ALUMINUM BOX. MOUNT L1 SECURELY TO CIRCUIT BOARD. IF L1 MOVES OR VIBRATES, THE FREQUENCY WILL SHIFT. BOTH 555 CHIPS CAN BE CMOS/LOW-POWER TYPES, BUT NOT ALL CMOS 555'S WILL WORK IN CIRCUIT. USE CIRCUIT FOR PAGING, REMOTE CONTROL, TRACKING, ANNOUNCING VISITORS, ETC. TO TRANSMIT LIGHT LEVEL OR TEMPERATURE AS A VARIABLE TONE, REPLACE R1 WITH PHOTORESISTOR OR THERMISTOR.

SPECIAL FCC RULE

THE FCC REQUIRES THAT "..THE DURATION OF EACH TRANSMISSION SHALL NOT BE GREATER THAN ONE SECOND AND THE SILENT PERIOD BETWEEN TRANSMISSIONS SHALL BE AT LEAST 30 TIMES THE TRANSMISSION DURATION BUT IN NO CASE LESS THAN 10 SECONDS." (47 CFR 15.122) WITH THE VALUES FOR R3, R4 AND C2 GIVEN HERE, THIS CIRCUIT FULFILLS THIS RULE. SEE NEXT PAGE FOR ADDITIONAL RULES.

FCC REGULATIONS

FCC RULES YOU SHOULD KNOW ABOUT INCLUDE.

1 EAVESDROPPING IS PROHIBITED.

2 A LOW-POWER TRANSMITTER THAT INTERFERES WITH RADIO OR TELEVISION RECEPTION MUST NOT BE OPERATED.

3. REQUIRED HOME-BUILT TRANSMITTER LABEL.

> I HAVE CONSTRUCTED THIS DEVICE FOR MY OWN USE. I HAVE TESTED IT AND CERTIFY THAT IT COMPLIES WITH THE APPLICABLE REGULATIONS OF FCC RULES PART 15. A COPY OF MY MEASUREMENTS IS IN MY POSSESSION AND IS AVAILABLE FOR INSPECTION.
> SIGNED: ____________ DATE: ______

ADDITIONAL RULES GIVE PERMISSIBLE SIGNAL STRENGTHS AND OTHER RESTRICTIONS. SEE 47 CFR, PART 15 FOR DETAILS (WRITE TO THE SUPERINTENDENT OF DOCUMENTS, U.S. GOVERNMENT PRINTING OFFICE, WASHINGTON, DC 20402).

GOING FURTHER

RADIO SHACK SELLS EASILY ASSEMBLED TRANSMITTER AND RECEIVER KITS. RADIO SHACK ALSO SELLS A WIDE RANGE OF CB EQUIPMENT. BOOKS ABOUT RADIO COMMUNICATIONS CAN BE FOUND AT MOST LIBRARIES. _POPULAR COMMUNICATIONS_, _73_, _QST_ AND _CQ_ ARE SOME OF THE MAGAZINES DEVOTED TO THE SUBJECT.

PROBABLY THE BEST GUIDE TO AMATEUR RADIO IS "THE ARRL HANDBOOK FOR THE RADIO AMATEUR." THIS ALL-INCLUSIVE BOOK, WHICH IS REVISED EACH YEAR, IS AVAILABLE FROM THE AMERICAN RADIO RELAY LEAGUE (NEWINGTON, CT 06111).

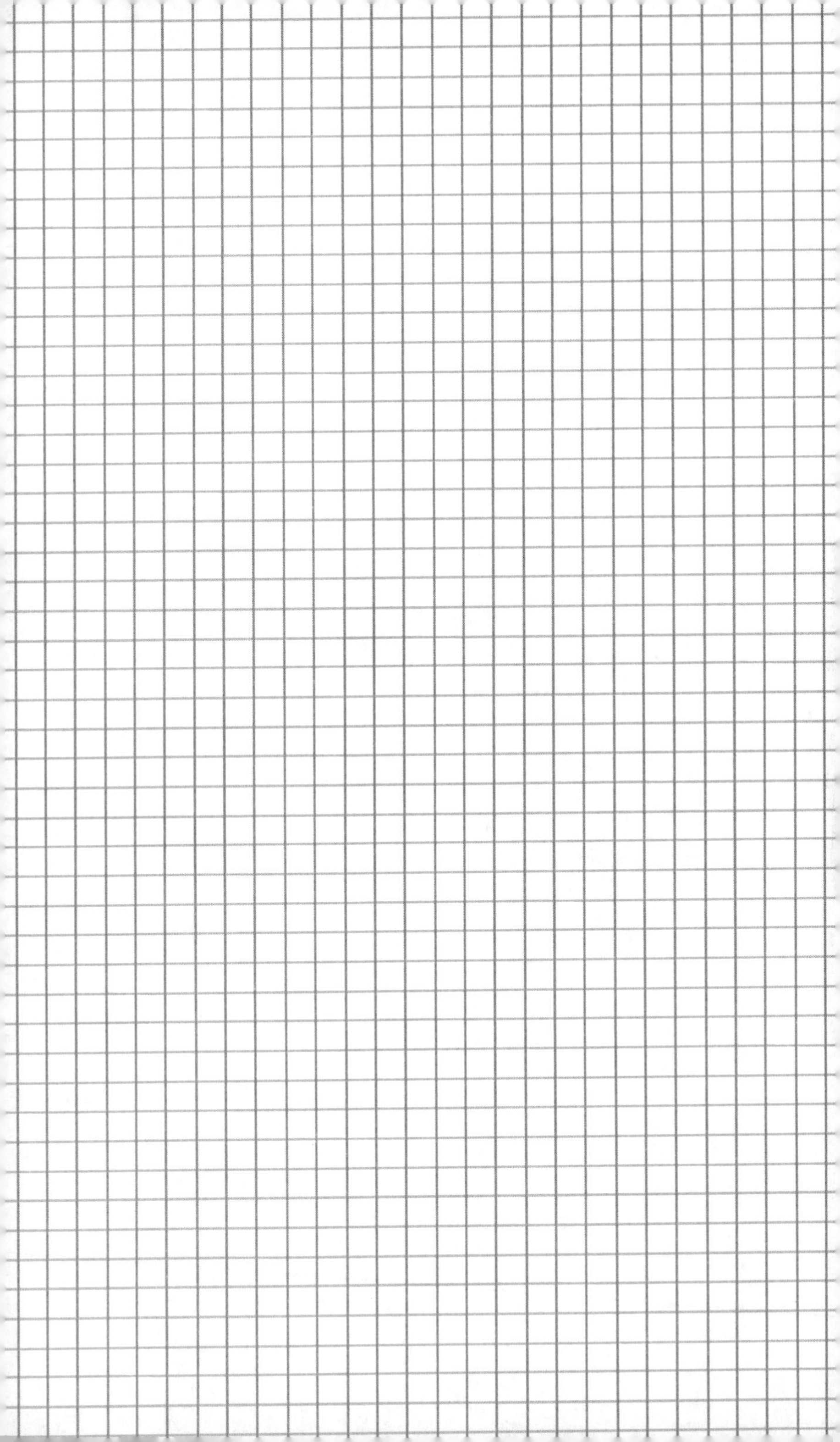

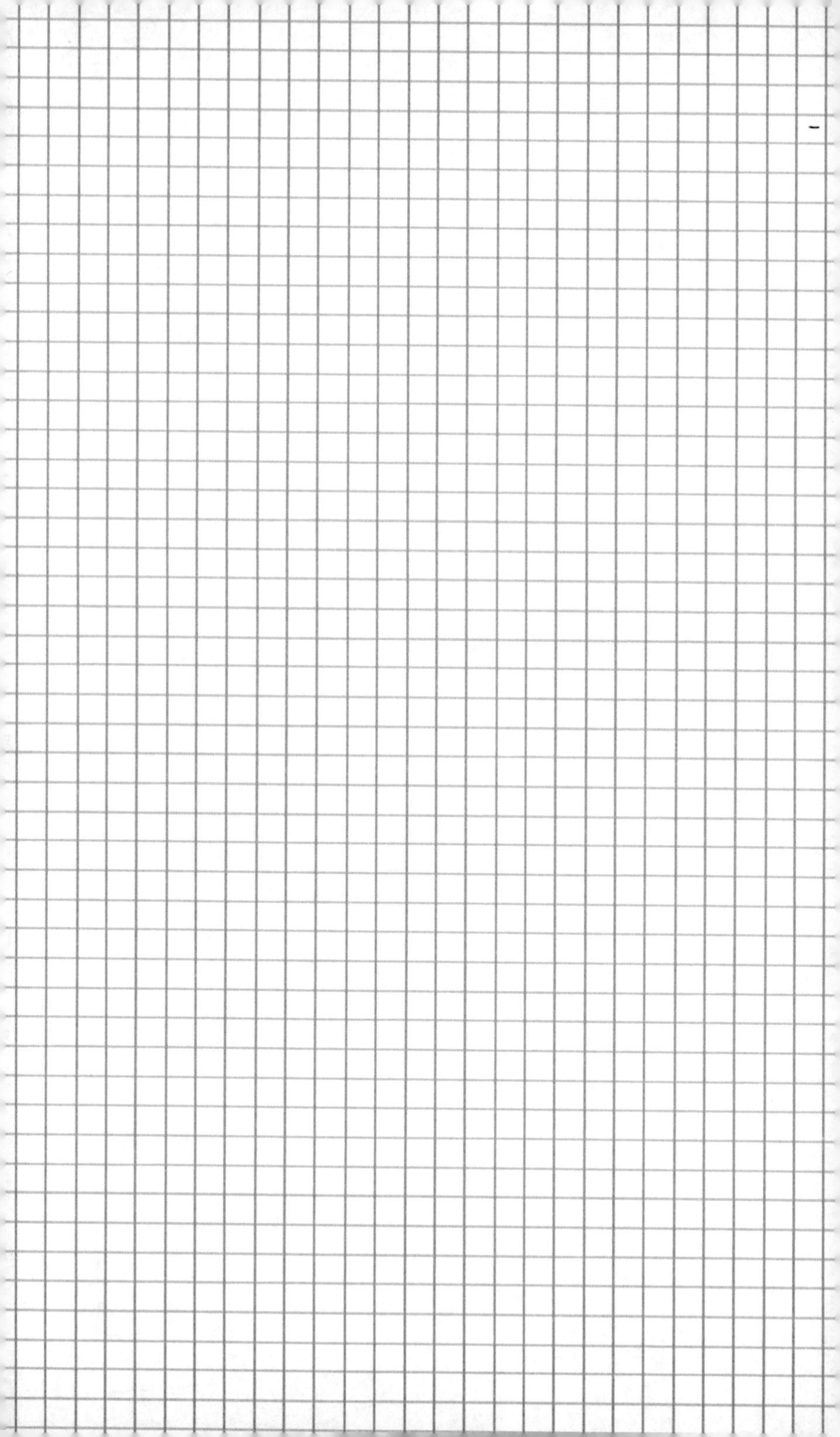

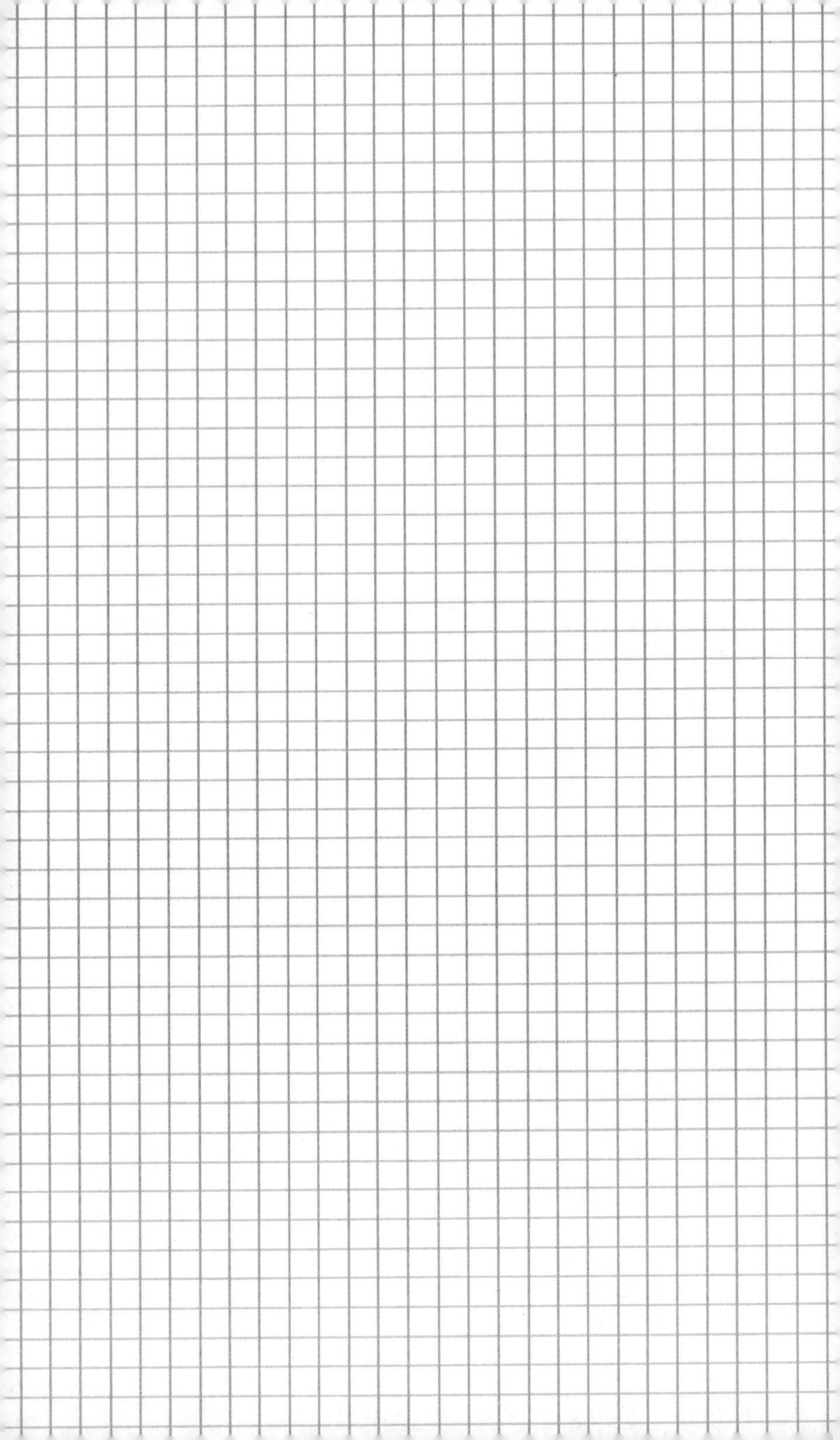

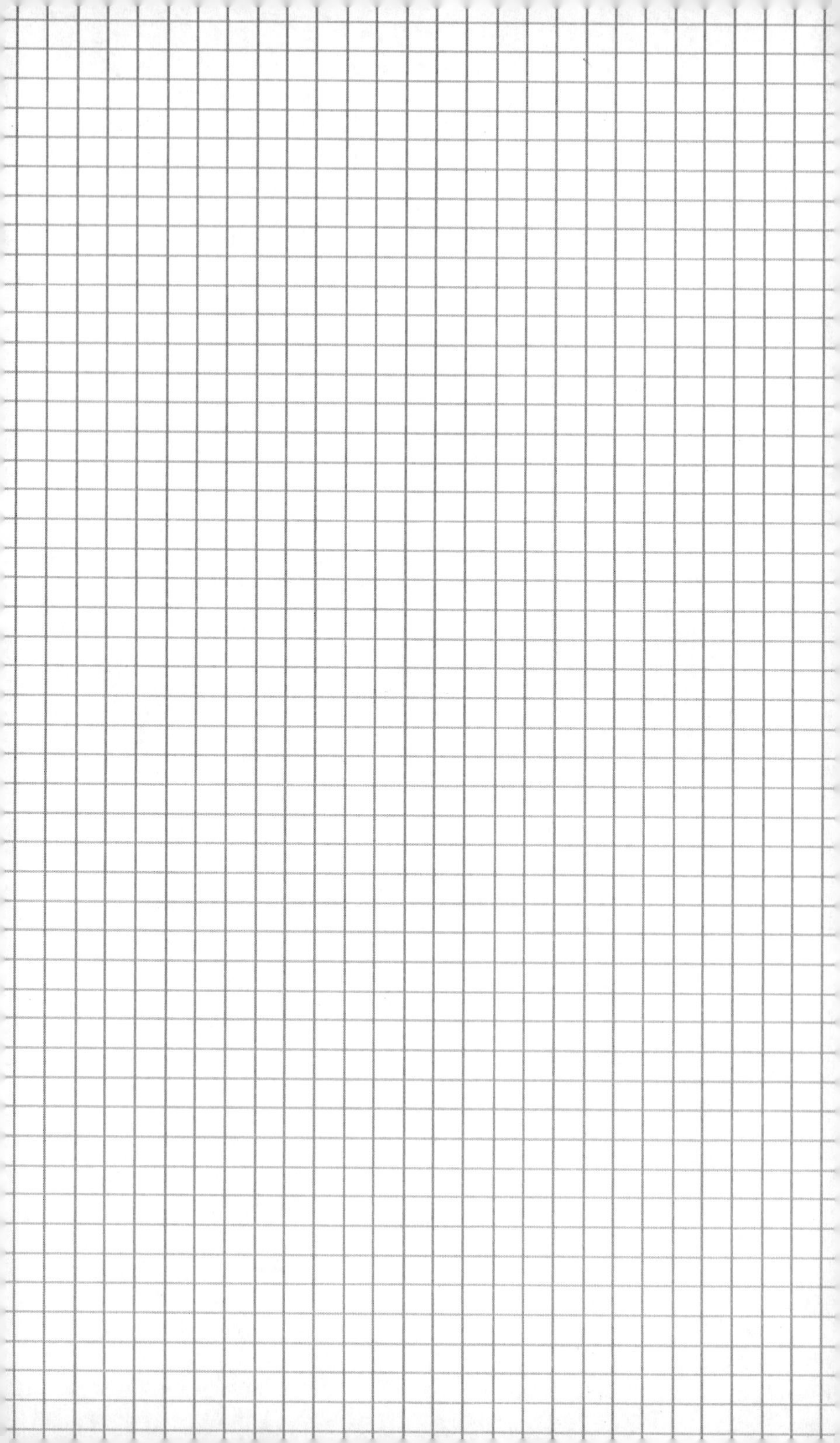